Electron Microscopy in Viral Diagnosis

Authors

Erskine L. Palmer, Ph.D.
Supervisory Research Microbiologist
Special Pathogens Branch
Division of Viral Diseases
Centers for Disease Control
Atlanta, Georgia

Mary Lane Martin
Research Microbiologist
Special Pathogens Branch
Division of Viral Diseases
Centers for Disease Control
Atlanta, Georgia

CRC Press
Taylor & Francis Group
Boca Raton London New York

CRC Press is an imprint of the
Taylor & Francis Group, an **informa** business

st published 1988 by CRC Press
ylor & Francis Group
00 Broken Sound Parkway NW, Suite 300
ca Raton, FL 33487-2742

issued 2018 by CRC Press

1988 by CRC Press, Inc.
RC Press is an imprint of Taylor & Francis Group, an Informa business

) claim to original U.S. Government works

brary of Congress Cataloging-in-Publication Data

lmer, Erskine L.
Electron microscopy in viral diagnosis.

Includes bibliographies and index.
1. Viruses--Identification.   2. Electron microscopy.
Diagnosis, Electron microscopic.   I. Martin, Mary Lane.
  Title   [DNLM: 1. Microscopy, Electron--methods.
Virus Diseases--diagnosis.   3. Virus Diseases--
terinaryc.   SF 780.4 P173e]
R387.P35   1988   616.9'250758   87-27669
BN 0-8493-4747-5

Library of Congress record exists under LC control number: 87027669

BN 13: 978-1-315-89254-2 (hbk)
BN 13: 978-1-351-07164-2 (ebk)

sit the Taylor & Francis Web site at http://www.taylorandfrancis.com and the

THE AUTHORS

Erskine Palmer received his B.S. and M.S. degrees from Florida State University and his Ph.D degree from the University of Mississippi. He has worked as a Research Virologist at the Centers for Disease Control since 1964 and is currently in charge of the Virology Division's Electron Microscopy Laboratory. He has conducted research on a wide variety of viruses of public health importance.

Mary Lane Martin holds a B.S. degree in Biology from Stetson University and a M.S. degree in biology from Georgia State University. She has worked as a Research Virologist at the Centers for Disease Control since 1966 and has carried out numerous studies on the morphology of mammalian viruses. She currently works with Dr. Palmer in the Special Pathogens Branch of the CDC.

TABLE OF CONTENTS

INTRODUCTION

Viruses are submicroscopic obligatory intracellular parasites that lack energy-generating enzyme systems for independent replication. A complete infectious virus particle, or virion, is composed of a genome of either RNA or DNA surrounded by a protein shell or membranous envelope which protects the genetic material from the environment and allows the virion to pass from one host cell to another. Outside of the cell viruses are inert, but might be considered "live" when viral nucleic acid enters the cell and causes synthesis of virus specific protein and nucleic acid. Thus viruses could be considered as either very simple microbes or complex chemicals.

Viruses are known which infect animals, plants, bacteria, algae, and fungi. This book will deal with animal viruses, primarily those of public health importance.

Classification of viruses is the responsibility of the International Committee on the Taxonomy of Viruses (ICTV). Most viruses have now been placed into virus families, genera, and species. The primary criterion for classification is morphology. Most of the animal viruses fit into about 20 morphological patterns. Another criterion is nucleic acid type. This can be either DNA or RNA. Viruses with the same morphology have the same type of nucleic acid. For example, all picornaviruses have single stranded RNA genomes and all adenoviruses have double stranded DNA genomes. Viruses within species also tend to have similar protein patterns discerned by polyacrylamide gel electrophoresis. Antigenic and genomic relatedness are other important factors for assigning a virus to a particular family or species. Within each virus family there are viruses which are related to varying degrees. This is one basis of ordering viruses of the major families into genera and species. A goal of the ICTV is to have a partly Latinized nomenclature for viruses which has international meaning. In this way a virus family has become a group of genera with common characters and the ending of the name of a viral family is " . . . viridae". Various virus genera are groups of species sharing certain common characters. The ending of the name of a viral genus is " . . . virus". Species are considered as a collection of viruses with like characters. Virus strains are different serotypes of the same species. Figure 1 is a line drawing of the major families of animal viruses arranged according to type of nucleic acid and size. The diagrams have been drawn to give an indication of the relative shape and size of the viruses but dimensions and shapes cannot be exact because some viruses are pleomorphic. The general characteristics of the major virus families are presented in Table 1. Details of virus taxonomy are updated on a timely basis by the ICTV in *Intervirology*.

Electron microscopy has been used to identify viruses since the late 1940s. In 1948 Nagler and Rake and Van Rooyen and Scott first showed by electron microscopy (EM) that there were differences in the morphology of virus particles in crust and vesicular fluids of lesions from patients with smallpox or chickenpox (varicella). Over two decades later the electron microscope was the primary diagnostic tool used during the global eradication of smallpox. Identification of most viruses by EM is now routinely accomplished in laboratories worldwide. Some of the diseases for which direct transmission EM can be utilized for rapid diagnosis include poxvirus infections, herpesvirus infections, nonbacterial gastroenteritises, hepatitis B, warts, some respiratory virus infections, parvovirus red cell aplasia, and some diseases of the brain.

One of the most useful methods for visualization of viruses is negative stain EM. The technique is well established and has become the method of choice for the rapid identification of viruses in clinical specimens. EM by negative stain methods is also a valuable adjunct to conventional isolation techniques because viruses in tissue culture fluids at a concentration of about 10^6 particles/mℓ can be grouped by morphology. Moreover, negative stain methods combined with immuno EM (IEM) allows visualization of virus in clinical specimens or culture fluids as low as 10^2 to 10^3 particles/mℓ. Thus, the contribution of IEM to virology

RNA VIRUSES

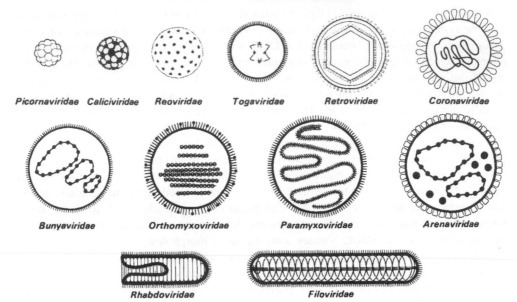

DNA VIRUSES

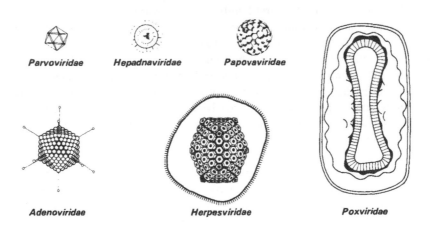

FIGURE 1. Stylized drawings of mammalian RNA and DNA virus families. The size, shape, and basic characteristic of each virus group is delineated. Internal components of enveloped viruses are also shown.

has enormous potential and the spectrum of viruses which may eventually be rapidly identified by this method has become limitless.

Viruses may also be identified by thin-section EM. Although not a rapid method, thin-section EM allows identification by replicative site as well as by morphology and is invaluable in situations when recovery of virus from tissue is difficult or hazardous. For example, thin section EM studies of glutaraldehyde fixed T4 lymphocytes played an important role in the identification of the retrovirus which causes the acquired immunodeficiency syndrome (AIDS). Negative stain EM played a key role in the identification of two other of the most recently

Table 1
GENERAL CHARACTERISTICS OF THE MAJOR VIRUS FAMILIES

Virus family	Nucleic acid type	Morphologic type	Size of capsid (nm)	Number of capsomeres	Size of enveloped virus (nm)	Special features	Examples of viruses
Picornaviridae	ssRNA	Naked icosahedron	20—30	32	—	+ stranded RNA	Poliovirus, rhinovirus, coxsackievirus, enterovirus, hepatitis A
Caliciviridae	ssRNA	Naked icosahedron	35—40	32	—	+ stranded RNA	Feline calicivirus, gastroenteritis virus
Reoviridae	dsRNA	Naked icosahedron	60—80	132	—	10—12 segments of RNP; capsomeres in a skew pattern	Reovirus, rotavirus, orbivirus
Togaviridae	ssRNA	Enveloped cubic capsid	25	—	70	+ stranded RNA	Eastern, Western encephalitis viruses
Flaviviridae	ssRNA	Enveloped cubic capsid	25	—	40—50	+ stranded RNA	Yellow fever, St. Louis encephalitis virus
Retroviridae	ssRNA	Enveloped cubic capsid	30—45	—	100—120	RNA dependent DNA polymerase	Human T cell leukemia/lymphotropic virus I, II, Human immunodeficiency virus
Bunyaviridae	ssRNA	Enveloped helical capsid	—	—	90—100	3 circular segments of RNP	Hantaan virus, LaCrosse virus
Orthomyxoviridae	ssRNA	Enveloped helical capsid	—	—	80—120	8 segments of RNP	Influenza A, B
Paramyxoviridae	ssRNA	Enveloped helical capsid	—	—	150—1000	"Herringbone" pattern of RNP; very pleomorphic	Mumps, measles, parainfluenza viruses
Coronaviridae	ssRNA	Enveloped helical capsid	—	—	75—160	+ stranded RNA	Human OC43 coronavirus
Arenaviridae	ssRNA	Enveloped helical capsid	—	—	50—300	2 circular segments of RNP	Lassa, Tacaribe complex viruses
Rhabdoviridae	ssRNA	Enveloped helical capsid	—	—	130—380 × 50—95	Bullet shaped	Rabies
Filoviridae	ssRNA	Enveloped helical capsid	—	—	60—70 wide	Long branching forms common	Marburg, Ebola
Parvoviridae	ssDNA	Naked icosahedron	18—26	32	—	Small size	Adeno-associated virus
Papovaviridae	dsDNA	Naked icosahedron	45—55	72	—	Capsomeres in a skew pattern; circular DNA	BK, JC, SV40, human wart viruses
Adenoviridae	dsDNA	Naked icosahedron	70—90	252	—	—	Adenoviruses 1—41 types
Herpesviridae	dsDNA	Enveloped cubic capsid	90—100	162	120—150	—	Herpes simplex, varicella-zoster, cytomegalo viruses, human B cell leukemia virus
Hepadnaviridae	dsDNA	Complex	—	—	40	Yields surface component 22 nm in diameter	Hepatitis B
Poxviridae	dsDNA	Complex	—	—	300—450 × 170—260	Brick shaped	Variola, Vaccinia
Iridoviridae	dsDNA	Complex icosahedral	—	—	130—300	Largest of isometric viruses	African swine fever virus

discovered virus groups, Marburg/Ebola viruses and rotavirus. Electron microscopy showed early on that the morphology of these viruses was unlike that of any virus group known at that time. It is likely that there are other morphological groups as yet undiscovered which will in due time be uncovered as a result of medical research.

Identification of viruses by electron microscopy is practicable because each of the currently known virus groups is composed of viruses with a distinct morphology. Members of each virus genus have structural characteristics that are recognizable when seen by EM, and atlases of virology are available for reference diagnosis of common and uncommon viruses.

Chapter 1

THE STRUCTURE OF ANIMAL VIRUSES

Two basic structural types of animal viruses were described in the early 1960s by Caspar and Klug: (1) icosahedral structure of isometric viruses and (2) helical structure. All known animal viruses, except poxviruses, belong to one of these two structural types. The poxviruses have a complex brick-like structure.

Icosahedral structure is also known as 5:3:2 symmetry because it possesses axes of 5-fold, 3-fold, and 2-fold symmetry (Figure 2). Icosahedral structure is found in both DNA and RNA viruses. Further, in some virus groups the icosahedral structure exists as a naked nucleocapsid, and in others, it is surrounded by an envelope studded with projections.

The morphological units, or capsomeres, of icosahedral viruses are arranged in such a way that the result is a rigid geometric configuration, an icosahedron, having 20 triangular sides and 12 five-fold vertices. Capsomeres may be arranged in straight rows between the vertices, an arrangement which often gives the capsid a hexagonal appearance when seen by EM. This structural class is known as P = 1 and is exemplified by the adenovirus group (Figure 3). Other structural classes of icosahedral symmetry are described as P = 3 and P = 7. In the P = 3 class, capsomeres are arranged in a zig-zag pattern between vertices. In the P = 7 class, neighboring vertices are set at an angle to each other. This results in a skew arrangement, and viruses belonging to this structural class may be *dextro* (skewed to the right) or *levo* (skewed to the left). The papovaviruses are structured as P = 7.

The capsomeres which make up the capsid shell are arranged in clusters with five- and six-fold symmetry. Capsomeres located on the vertices of the capsid are made up of a group of five subunits and are called pentameres, whereas capsomeres located between the vertices are made up of six subunits and are called hexameres. The simplest possible structure of this type would be composed of 60 subunits comprising 12 pentameres located at the vertices of the icosahedron and would have no hexameres. Larger isometric viruses have 12 pentameres and also hexameres, the number of which is always a multiple of 10.

The structure of icosahedral viruses may be described by their triangulation number, or T-number. T-number is defined as the number of equilateral subtriangles into which each of the 20 sides of the icosahedron can be divided, each subtriangle containing three structural subunits. One way the T-number may be calculated is by counting the number (n) of hexameres between two neighboring pentameres and using the equation $(n + 1)^2 = T$. For example, the herpesvirus capsid has 3 hexameres between neighboring pentameres, so its T number is calculated as T = 16 (Figure 4). The T-number is indicative of the number of subunits and can be used to calculate the total number of capsomeres (N) by the equations N = 10 T + 2 or N = 10(T − 1) hexameres + 12 pentameres. Thus, the herpesvirus nucleocapsid, which has a T number of 16, has 162 capsomeres. Some icosahedral viruses such as the Reoviridae have a capsid formed by sharing of subunits which results in a prominent pattern of holes. Nevertheless, the arrangement of the subunits is consistent with icosahedral symmetry. The icosahedral viruses have a relatively constant size and shape within each genus. An example of the icosahedral viruses is poliovirus (Figure 5).

The capsid shell of icosahedral viruses surrounds the nucleic acid core of the virus, and the capsid and enclosed genome are referred to as the nucleocapsid. This term is also used to describe the helical arrangement of protein and nucleic acid found in the other major structural group, the helical viruses.

The helical viruses contain RNA which is assembled with protein subunits into a helical nucleocapsid with the RNA located in a channel in the center of the helix. The single axis of symmetry passes longitudinally down this helical center (Figure 6). The nucleocapsid of different virus groups varies in the periodicity of its coils, so that some are very rigid and

FIGURE 2.   Drawings of icosahedra on axes of 5-, 3-, and 2-fold symmetry.

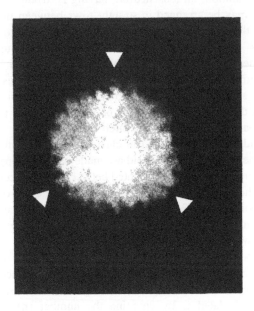 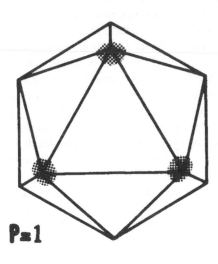

FIGURE 3.   Adenovirus and a drawing representing the structural class P = 1. The capsomeres of adenovirus are arranged in a straight line between the 5-fold vertices giving the capsid a hexagonal appearance. A triangular facet is indicated by arrows.

others are more flexible rods (Figures 7A, 7B). The nucleocapsids of viruses of the Bunyaviridae and Arenaviridae are flexible circular structures (Figure 8). The nucleocapsid is coiled further to fit into a pleomorphic lipid-containing envelope which is covered with characteristic surface projections, or spikes, which are composed of glycoprotein. The surface projections seen on the surface of influenza virus (Figure 9) are typical structures of this kind. Whereas the icosahedral viruses have a relatively constant size and shape within each family, the helical viruses tend to be heterogeneous in size and are frequently pleomorphic. By EM, these may appear as generally spherical particles such as the coronaviruses (Figure 10), or as more defined rod-shaped particles such as the rhabdoviruses (Figure 11). Pleomorphism is common in the orthomyxovirus group (Figure 12), and virions may be seen as spherical particles of widely varying sizes, as bizarre ameboid shapes, or as filaments that may be several micrometers in length.

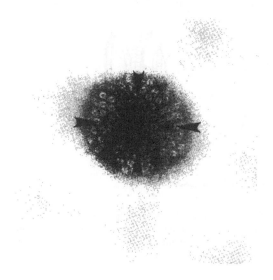

FIGURE 4. A degenerating capsid of herpes simplex virus. The pentagonal capsomeres are missing (arrows) and the particle is lying on an axis of 2-fold symmetry. Capsomeres are seen arranged in rows. There are three hexameres between two neighboring pentameres. The T number calculated by the equation $(n + 1)^2 = T$ is therefore 16. There are 162 capsomeres. UA stain $\times$ 293,000.

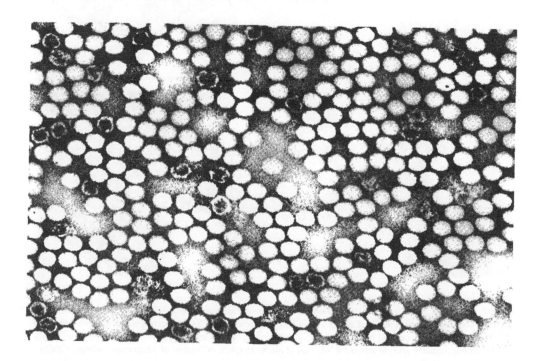

FIGURE 5. Poliovirus, an *Enterovirus* species representative of small icosahedral viruses with a relatively constant size and shape. PTA stain $\times$ 117,500. (Courtesy of Joseph Esposito, CDC.)

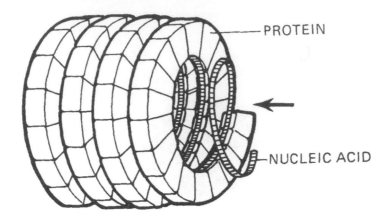

FIGURE 6.   Stylized drawing illustrating helical symmetry. The single rotational axis is indicated by an arrow. Protein subunits surround the nucleic acid, a portion of which is shown free in the drawing. The protein subunits of each turn of the helix are offset in relation to those of the previous turn.

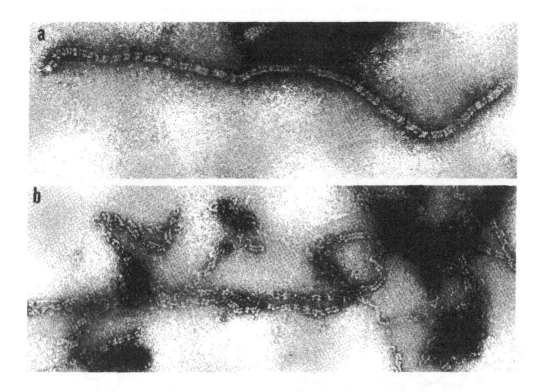

FIGURE 7.   (A) Helical nucleocapsid of a paramyxovirus showing an example of a rigidly structured nucleocapsid. The helix is continuous with a width of about 18 nm and the nucleocapsid has a "herringbone" pattern. UA stain × 155,916; (B) The nucleocapsid of rabies virus, a rhabdovirus, which appears flexible when lying free. It is a helix with protein radially arrayed around RNA. UA stain × 155,916. (Courtesy of Makonnen Fekadu, CDC.)

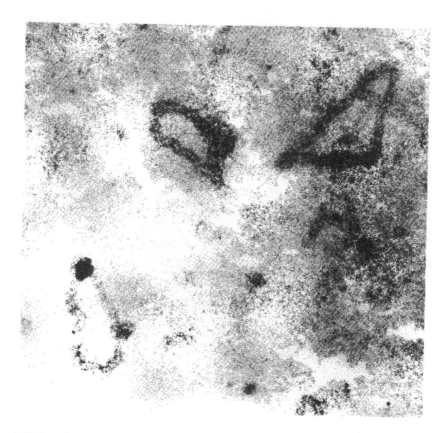

FIGURE 8.   Nucleocapsids of the bunyavirus LaCrosse. The structures are circular and composed of three size classes. The symmetry of these structures is probably helical. UA stain × 173,400.

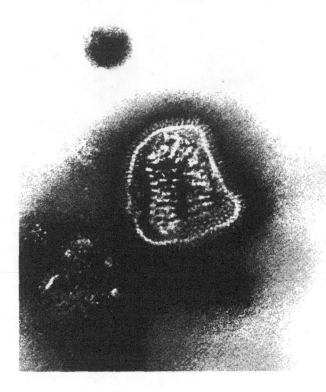

FIGURE 9.   A roughly spherical form of influenza virus stained with UA. Spikes are visible on the surface. The stain has penetrated the envelope and the helical nucleocapsid can be seen as parallel bars × 156,260.

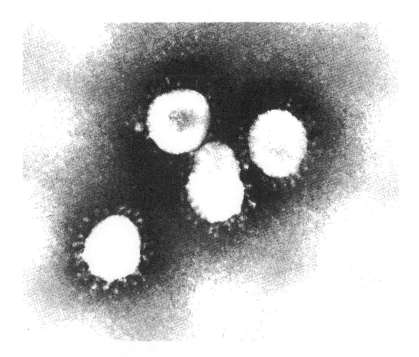

FIGURE 10.   Coronavirus OC43, an enveloped virus with a spherical appearance. PTA stain × 148,500. (Courtesy of Frederick Murphy, CDC.)

FIGURE 11. Vesicular stomatitis virus, a rhabdovirus, with a relatively rigid "bullet" shape. PTA stain × 105,100.

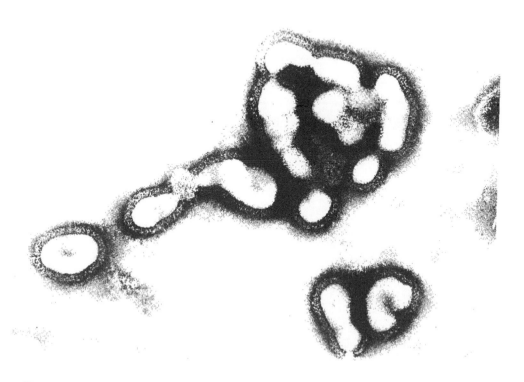

FIGURE 12. Influenza virus. A helical virus which often appears pleomorphic. UA stain × 136,728.

Chapter 2

NEGATIVE STAIN EM

The technique of negative staining, developed by Brenner and Horne in 1959, utilizes an electron opaque substance to surround the virus or other biological structure, giving contrast between the electron-lucent biological material and the background against which it is viewed. The negative stain not only surrounds the virus particle but penetrates between the viral structural units and even into the interior of the virus to delineate morphological features. The areas not penetrated by stain appear electron-lucent against a contrasting opaque background. Image formation is the result of electrons being absorbed or deflected by the stain (Figure 13). A good negative stain must form a nongranular matrix upon drying and have the ability to penetrate into the tiny crevices of biological material. The substances commonly used as negative stains are the salts of heavy metals — primarily 2 to 4% sodium or potassium phosphotungstate (PTA) and 0.5% uranyl acetate (UA). These can be used as aqueous solutions in the acid pH range. PTA is usually adjusted to a pH of around 7 by the addition of either sodium hydroxide or potassium hydroxide, whereas UA is used at a pH of about 4.5 without adjustment. UA may act as either a positive or negative stain because it adsorbs to protein and nucleic acid, and it is therefore a particularly useful stain for locating viral ribonucleoprotein. Silicotungstic acid, ammonium molybdate, uranyl formate, nitrate, or oxalate are also suitable as negative stains but their use is not as widespread as that of PTA and UA.

When interpreting images from negatively stained preparations, one must keep in mind that the images may be "two-sided"; that is, the image of the upper half of the particles may be superimposed upon that of the lower half of the particles, which is more completely submerged in the stain matrix. In fact, much of the image information may be contributed by the lower portion of the particle since that is where the stain is actually "pooled" around the particle. When the pseudoreplica technique (see below) is used for specimen preparation, however, the lower portion of the virus is encased in plastic and is relatively inpenetrable by stain so that the problem of superimposition is less severe.

There are many techniques of negative stain EM, but there is no universally accepted method for grid preparation. A general method for detecting virus in liquid preparations is to mix equal volumes of PTA and virus suspension and place a drop of the mixture onto a grid coated with carbon or with a plastic film such as Formvar or Parlodion. Excess fluid is removed by touching the edge of the grid with filter paper and letting the fluid air-dry. Other methods are (1) to let the virus air-dry on a grid before adding stain, and (2) float a grid on a drop of the stain-virus suspension, or on virus alone, then on a drop of stain. These methods are sufficient to detect virus at a concentration of about $10^{5.5}$ to 10^6 particles/mℓ. However, at this concentration only an occasional particle may be seen on a few grid squares.

There are several methods of desalting virus suspensions before preparing grids. One of these is the agar diffusion technique. In this technique a drop of virus suspension is placed on the surface of agar on a microscope slide, in a microtiter plate well, or some other convenient vehicle, and a grid is floated on the drop. As the liquid dries, salts diffuse into the agar and larger material remains on the grid. These are then negatively stained. Alternately, the liquid is dried on an agar surface and the grid placed over the dried material. Virus and other material on the surface of the agar adhere to the grid. Salts can also be removed by washing of grids after air-drying.

The pseudoreplica technique removes salts and concentrates samples simultaneously. The method takes a few minutes to complete, but the resulting grids contain a better representation of what is in a specimen than do the simpler methods. In this procedure, a drop of virus suspension is allowed to dry on a small block of 2% agarose. Placing the agar block on one

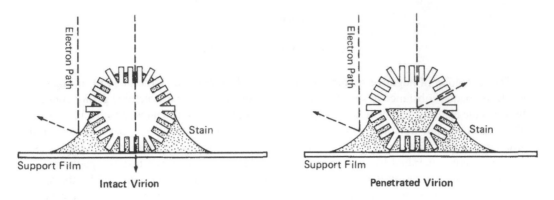

FIGURE 13.   Diagram showing how images are formed as a result of electrons being adsorbed or deflected by negative stains.

corner of a microscope slide facilitates handling the specimen during subsequent steps. Salts diffuse into the agar while virus remains dried on the agar surface. The dried material is covered with a drop of 0.5% Formvar which forms a membrane film as it is drained off. The Formvar picks up most of the material that is left on the agar surface after drying. The Formvar film is then floated off of the agar block onto the surface of a negative stain contained in a small petri dish. A grid is placed onto the floating film, and both grid and adherent film are picked up with a small peg. There are no problems with salt artifacts so aqueous UA, which is precipitated by electrolytes, can be used as a positive and negative stain. The pseudoreplica technique also allows monitoring of fractions from density gradients without prior dialysis. In addition, three-dimensional viral structure can sometimes be delineated by this technique (Figure 14) when particles are oriented so that the view is directed toward the inside of tubular structures. A diagram of the pseudoreplica technique is shown in Figure 15. Viruses which can be seen relatively easily in body fluids by direct EM are listed in Table 2.

Prefixation of liquid preparations of some viruses with glutaraldehyde prior to negative staining can be done to accentuate features without distorting viral structure. This is particularly true of viruses of the family Bunyaviridae and some members of the Togaviridae. The surface of Bunyamwera viruses looks like a tiny fringe when particles are unfixed (Figure 16) and like knobs when fixed with glutaraldehyde and stained with UA. Glutaraldehyde fixation and UA staining allowed von Bonsdorf and Pettersson to discern an icosahedral pattern to the surface spikes of the Bunyavirus, Uukuniemi virus. This method has been used to show that the surface structure of viruses of each the four recognized genera of the Bunyaviridae are different. The causal agent of Korean hemorrhagic fever (Hantaan virus) and serologically related viruses were also found to have a surface structure not previously described for animal viruses. Micrographs of these viruses are shown in Figures 17A, 17B, and 17C.

Other viruses show a slight degree of distortion after glutaraldehyde fixation. Influenza virus spikes are somewhat disorganized (Figure 18) and rabies virus particles (Figure 19) tend to round up and lose some of the bullet-shaped structure common to the Rhabdoviridae. Still other viruses are significantly distorted by glutaraldehyde. Arenaviruses become "sticky" and form long chains and particles become difficult to recognize by EM (Figure 20).

Marburg and Ebola viruses (Figure 21) do not appear to be markedly affected by glutaraldehyde fixation. Surface spikes and ribonucleoprotein of these viruses can readily be discerned with fixed particles.

Glutaraldehyde fixation has allowed visualization of structures that were not discernible in unfixed preparations. Nevertheless care must be taken in the interpretation of micrographs of fixed particles because of the varying effects of fixation on different virus groups.

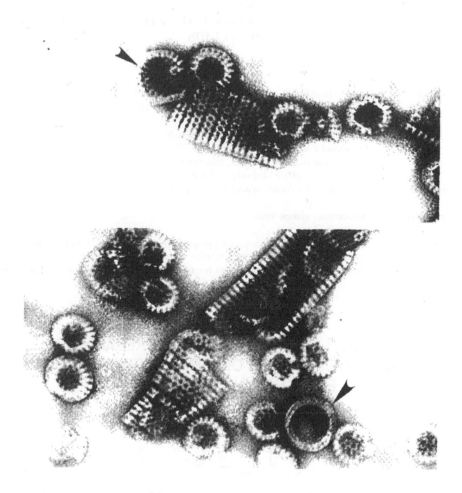

FIGURE 14.   Subviral forms of rotavirus showing 3-dimensional tubules (arrows) sometimes seen by negative staining of preparations prepared for EM by the pseudoreplica technique. UA stain × 195,325.

Table 2
THE VIRUSES DETECTABLE BY DIRECT NEGATIVE STAIN OF HUMAN BODY FLUIDS

Specimen	Viruses detectable by direct EM
Vesicular fluid	Herpes simplex 1, 2, varicella-zoster viruses, poxviruses
Urine	Papovavirus JC in immunocompromised patients, cytomegalovirus in patients with congenital cytomegalic inclusion disease
Serum	Hepatitis B, unclassified parvoviruses (aplastic anemia)
Tears	Occasionally adenovirus and enterovirus 70
Feces	Rotavirus, adenovirus, reovirus, calicivirus, coronavirus, astrovirus, small round viruses, bacteriophage, and others (Table 3)
Nasopharyngeal secretions	Occasionally respiratory syncytial virus, parainfluenza virus, coronavirus
Cerebrospinal fluid	Herpesvirus, mumps virus
Warts	Papilloma virus
Brain biopsy	BK papovavirus from patients with progressive multifocal leukoencephalopathy (including AIDS patients), herpes simplex virus from brain biopsies of patients with herpes encephalitis, measles virus from brain tissue of patients with subacute sclerosing panencephalitis

PSEUDOREPLICA TECHNIQUE FOR PREPARATION AND NEGATIVE STAINING OF VIRAL SPECIMENS FOR ELECTRON MICROSCOPY

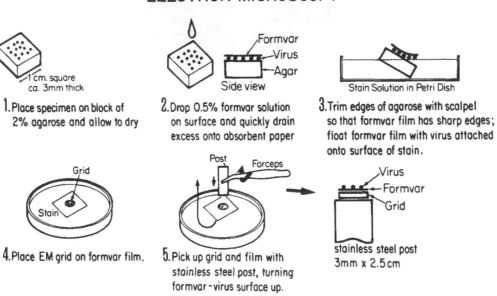

1. Place specimen on block of 2% agarose and allow to dry

2. Drop 0.5% formvar solution on surface and quickly drain excess onto absorbent paper

3. Trim edges of agarose with scalpel so that formvar film has sharp edges; float formvar film with virus attached onto surface of stain.

4. Place EM grid on formvar film.

5. Pick up grid and film with stainless steel post, turning formvar-virus surface up.

FIGURE 15.   Diagram of the pseudoreplica technique used to prepare specimens for negative stain EM.

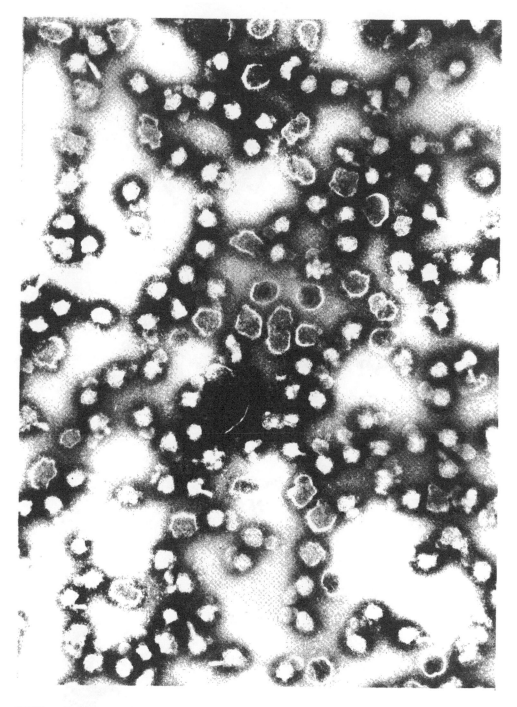

FIGURE 16. LaCrosse Virus, a *Bunyavirus*, unfixed and stained with PTA. The surface appears as a tiny fringe. UA stain × 74,925.

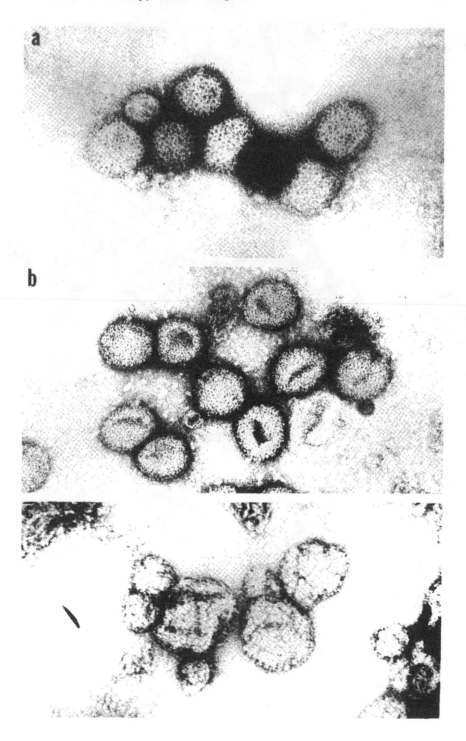

FIGURE 17.   Viruses of the family Bunyaviridae after glutaraldehyde fixation and negative staining with UA. (A) Rift Valley Fever Virus × 175,793; (B) Uukuniemi Virus × 136,728; (C) Tchoup-itoulas Virus × 153,360.

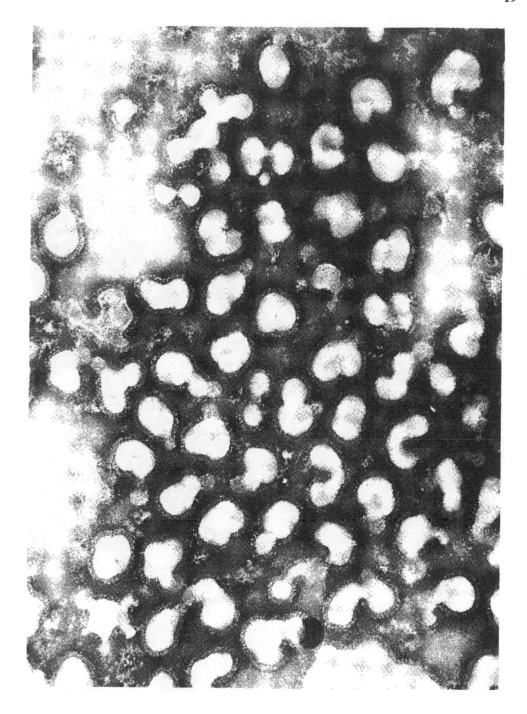

FIGURE 18.   Glutaraldehyde fixed influenza virus. Note that spikes are slightly disorganized, but that particles are still easily identifiable as influenza. UA stain × 117,195.

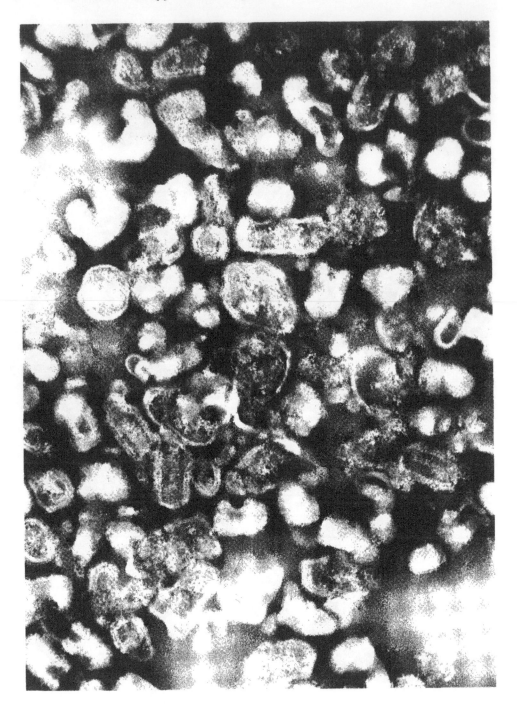

FIGURE 19.   Glutaraldehyde fixed rabies virus. Particles are generally more rounded than usually seen with Rhabdoviridae. Particles are still easily identifiable as rhabdoviruses. UA stain × 105,100.

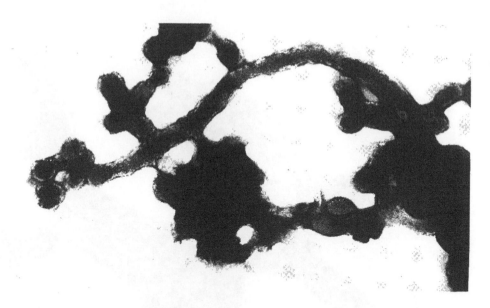

FIGURE 20. Glutaraldehyde fixed Tacaribe virus, an arenavirus. Particles are "sticky" and form long chains. Particles are not easily recognized as arenaviruses. UA stain × 89,460.

FIGURE 21.   Glutaraldehyde fixed Marburg virus. Particles do not appear distorted by glutaraldehyde. UA stain × 94,500.

REFERENCES

1. **Caspar, D. L. and Klug, A.,** Physical principles in the construction of regular viruses, *Cold Spring Harbor Symp. Quant. Biol.,* 27, 1, 1962.
2. **Dalton, A. J. and Haguenau, F.,** *Ultrastructure of Animal Viruses and Bacteriophages: An Atlas,* Academic Press, New York, 1973.
3. **Doane, F. W. and Anderson, N.,** *Electron Microscopy in Diagnostic Virology: A Practical Guide and Atlas,* Cambridge University Press, London, 1986.
4. **Kelen, A. E., Hathaway, A. E., and McLeod, D. A.,** Rapid detection of Australia/SH antigen and antibody by a simple and sensitive technique of immunoelectron microscopy, *Can. J. Microbiol.,* 17, 993, 1971.
5. **Martin, M. L., Lindsey-Regnery, H. L., Sasso, D. R., McCormick, J. B., and Palmer, E.,** Distinction between bunyaviridae genera by surface structure and comparison with Hantaan virus using negative stain electron microscopy, *Arch. Virol.,* 86, 17, 1985.
6. **Nagler, E. P. O. and Rake, G.,** Use of the electron microscope in diagnosis of variola, vaccinia and varicella, *J. Bacteriol.,* 55, 45, 1948.
7. **Palmer, E. L. and Martin, M. L.,** *An Atlas of Mammalian Viruses,* CRC Press, Boca Raton, Fla., 1982.
8. **Sharp, D. G.,** Sedimentation counting of particles via electron microscopy, in *Proc. 4th Int. Cong. Electron Microscopy,* Vol. 2, Springer-Verlag, Berlin, 1959, 542.
9. **Smith, K. O.,** Identification of viruses by electron microscopy, in *Methods in Cancer Research,* Busch, H., Ed., Academic Press, New York, 1967, 545.
10. **Van Rooyen, C. E. and Scott, G. G.,** Smallpox diagnosis with special reference to electron microscopy, *Can. J. Pub. Health,* 39, 467, 1948.
11. **Williams, R. C. and Fisher, H. W.,** *An Electron Microscopic Atlas of Viruses,* Charles C Thomas, Springfield, Ill., 1974.

REFERENCES FOR TABLE 2

1. **Cruickshank, J. G., Bedson, H. S., and Watson, D. H.,** Electron microscopy in the rapid diagnosis of smallpox, *Lancet* 2, 527, 1966.
2. **Dane, D. S., Cameron, C. H., and Briggs, A.,** Virus-like particles in serum of patients with Australian-antigen-associated hepatitis, *Lancet,* 1, 695, 1970.
3. **Doane, F. W., Anderson, N., and Chatiyanonda, K.,** Rapid laboratory diagnosis of paramyxovirus infections by electron microscopy, *Lancet,* 2, 751, 1967.
4. **Evans, A. S. and Melnick, J. L.,** Electron microscopic studies of the vesicle and spinal fluids from a case of Herpes-zoster, *Proc. Soc. Exp. Biol. Med.,* 71, 283, 1949.
5. **Harta-Barbosa, L., Fuccillo, D. A., Sever, J. L.,** Subacute sclerosing panencephalitis: isolation of measles virus from a brain biopsy, *Nature (London),* 221, 974, 1969.
6. **Katz, M., Oyanagi, S., and Koprowski, H.,** Subacute sclerosing panencephalitis: structures resembling myxovirus nucleocapsids in cells cultured from brain, *Nature (London),* 222, 888, 1969.
7. **LeCatsas, G., Prozesky, O., Van Wyk, L., and Els, H. J.,** Papovavirus in urine after renal transplant, *Nature (London),* 241, 343, 1973.
8. **Lee, F. K., Nahmias, A. J., and Stango, S.,** Rapid diagnosis of cytomegalovirus infection in infants by electron microscopy, *N. Engl. J. Med.,* 299, 1266, 1978.
9. **Nagington, J.,** Electron microscopy in differential diagnosis of pox virus infections, *Br. Med. J.,* 2, 1499, 1964.
10. **Payne, F. E., Baublis, J. V., and Itabashi, H. H.,** Isolation of measles virus from cell cultures of brain from a patient with subacute sclerosing panencephalitis, *N. Engl. J. Med.,* 281, 585, 1969.
11. **Peters, D., Nielsen, G., and Bayer, M. E.,** Reliability of the rapid EM diagnosis of smallpox, *Dtsch. Med. Wochenschr.,* 87, 2240, 1962.
12. **Takemoto, K. K., Barker, L. F., Pineda-Tamondong, C. F., and Nelson, D.,** Isolation of papovavirus from brain tumor and urine of a patient with Wiscott-Allrich syndrome, *J. Natl. Canc. Inst.,* 53, 1205, 1974.

Chapter 3

NEGATIVE STAIN IMMUNE ELECTRON MICROSCOPY

Immune electron microscopy (IEM) is the direct visualization of antigen and antibody complexes by negative stain or thin section EM. In virology, this includes visualization of viruses and antigenic subviral components.

Historically, this approach was reported in 1941 by Anderson and Stanley, when it was noted that tobacco mosaic virus appeared morphologically different when mixed with immune serum than when mixed with nonimmune serum. However, IEM was slow in developing because of tedious and inadequate methods, such as metal-shadow casting, of specimen preparation. Then in 1959, Brenner and Horne introduced negative staining, a rapid technique which made it possible to more easily visualize viruses. Virus-antibody complexes could also be seen by this technique, for which virus is first mixed with antiserum then negatively stained. The value of IEM was demonstrated soon after the introduction of negative staining procedures by demonstrating recognizable changes on the projections of influenza virus after incubation of the virus with specific antiserum. Individual antibody molecules attached to the virus were resolved. At about the same time a mixture of poliovirus and bacteriophage was used to show that each particle type was aggregated only by homologous antiserum.

Some uses of negative stain IEM are:

1. To identify elusive viruses which have been difficult to cultivate and otherwise may have gone undetected for perhaps years. Some of these are listed in Table 3 and some IEMs of these viruses are shown in Figures 22, 23, 24, and 25.
2. To aggregate low concentrations of virus into immune complexes so that the particles can be more easily seen. IEM generally concentrates virus 100- to more than 1000-fold in the area visualized.
3. To determine the antigenic relationships among newly discovered viruses such as the large number of virus isolates from fecal extracts of persons with nonbacterial gastroenteritis.
4. To serotype viruses. Papovaviruses and enteroviruses are examples. An immune aggregate of coxsackie B virus is shown in Figure 26. Purified virus can also be used to titer sera.
5. To aggregate viruses into immune complexes so that these can be differentiated from debris.
6. To study the attachment of immunoglobulins to specific binding sites on virus particles as was done with influenza virus.
7. To identify subviral components especially with the use of monoclonal antibody. IEM of single shelled rotavirus is shown in Figure 27. Virus was labeled with monoclonal antibody to the inner capsid. The antiserum reacts with single-shelled but not double-shelled particles.

There are a variety of methods used to prepare grids for IEM. With small volumes, one can mix 20 $\mu\ell$ virus and 20 $\mu\ell$ serum. These are incubated at 37°C 1 hr, then overnight at 4°C. Two grids are prepared by the pseudoreplica technique. One is stained with 2% PTA, pH 6.5—7.0, and the other with 0.5% aqueous UA, pH 4.5. When large volumes of virus suspension are available, it is advisable to mix larger volumes, depending upon the availability of serum and virus, then after incubation, centrifuge the aggregates into a pellet at about 10,000 × g for 20 to 30 min. The pellet can then be resuspended in a small amount of distilled water. Others prefer to use the drop-to-drop method of grid preparation, agar diffusion, or some version of the serum in agar method in which antiserum is incorporated

Table 3
SOME ELUSIVE VIRUSES DETECTED BY IEM

Virus	Source	Ref.
Papilloma	Wart material	1
Rubella	Tissue Culture	3
Hepatitis B (Dane particle and surface antigen)	Serum	2
Hepatitis A	Stool	4
Norwalk agent	Stool	5
Korean hemorrhagic fever (Hantaan)	Tissue culture	6

into agar in microtiter plate wells or on microscope slides. Grids are placed onto the agar and virus preparations are added on top of the grids. These are allowed to air dry and then negatively stained. Methods of IEM using antiserum-treated grids have also been devised. This solid-phase IEM method has several variations. Antibody can first be adsorbed to coated grids and used to trap virus particles. Alternately, a second layer (decorator) of antibody can be added to the trapped virus in a double antibody type technique. *Staphylococcus aureus* protein A has also been used as a binding agent for virus specific IgG followed by addition of virus, then, if desired, a decorator antibody.

Regardless of the method chosen to do IEM, the antisera used in the test needs to be chosen carefully. Some sera contain antigen-antibody complexes, others may contain viruses, and many have large amounts of virus-like lipid and protein debris. For research and reference diagnosis it is best to use isolated immunoglobulins or monoclonal antibody.

Sera with known titers obtained in established tests, such as neutralization, can be diluted for IEM one-half the endpoint titer in that test. Serum is heated to 56°C for 1 hr to inactivate complement, filtered through 0.22 μm filters, and clarified by centrifugation at 100,000 × *g* for 30 min. Before use, the serum is examined by negative stain EM with both PTA and UA to assure that it does not contain virus-like particles.

The concentration of serum antibody in relation to virus markedly affects the appearance of virus as seen by IEM. When antibody concentration is low, small aggregates are seen. These are usually composed of a few individual particles. As the antibody concentration increases, larger aggregates occur. At antibody excess, aggregates occur less frequently and individual particles may have a halo of antibody. Particle surface detail is also less defined (Figures 28, 29, 30, and 31).

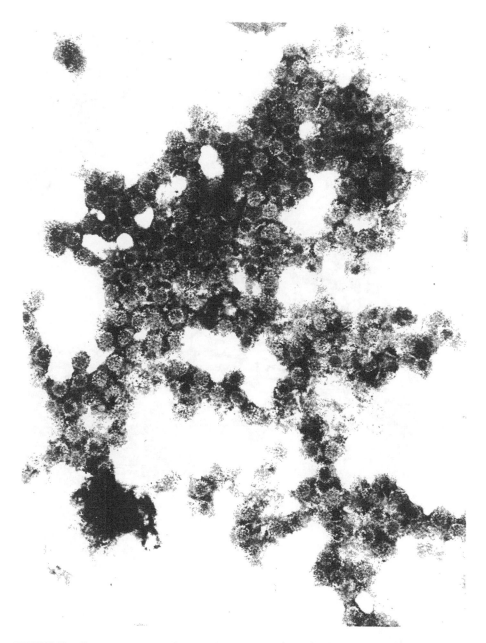

FIGURE 22.   Immune aggregate of papovavirus in culture fluid of infected monkey kidney cells. UA stain × 237,600.

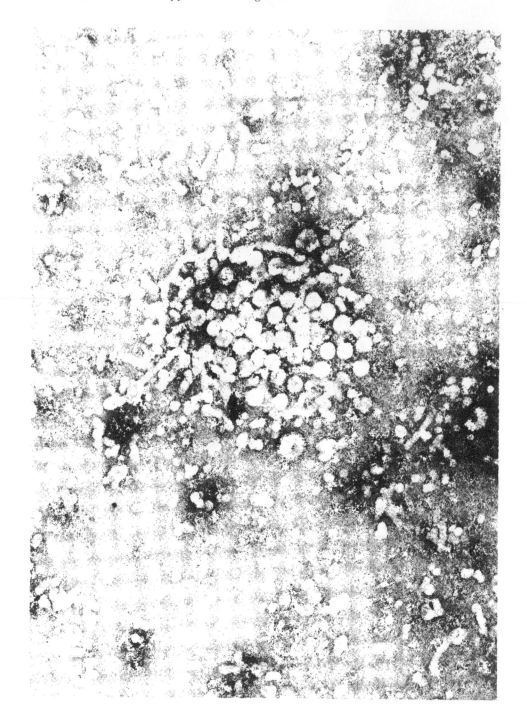

FIGURE 23.   Immune aggregate of hepatitis B in plasma from a patient with AIDS. PTA stain × 102,240.

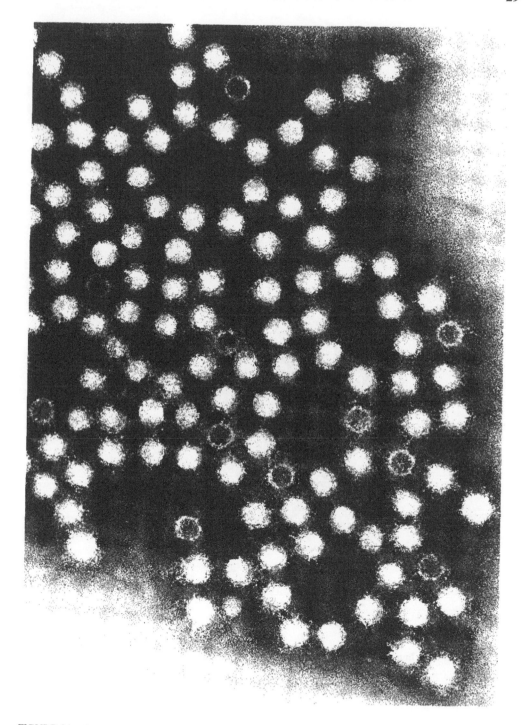

FIGURE 24.   Immune aggregate of hepatitis A virus derived from stool. PTA stain × 261,000. (Courtesy of Jim Cook, CDC.)

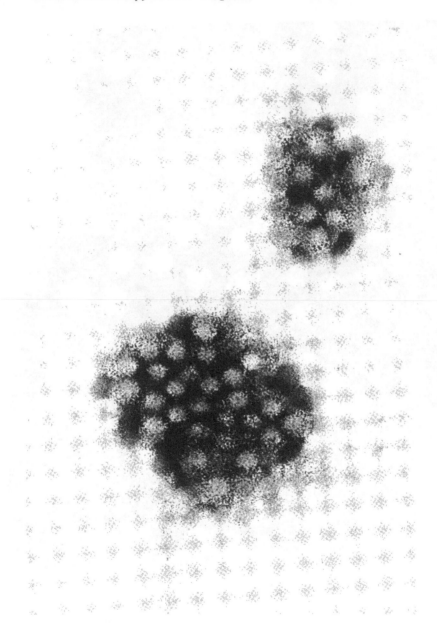

FIGURE 25.   Immune aggregate of Norwalk agent derived from stool. UA stain × 195,325.

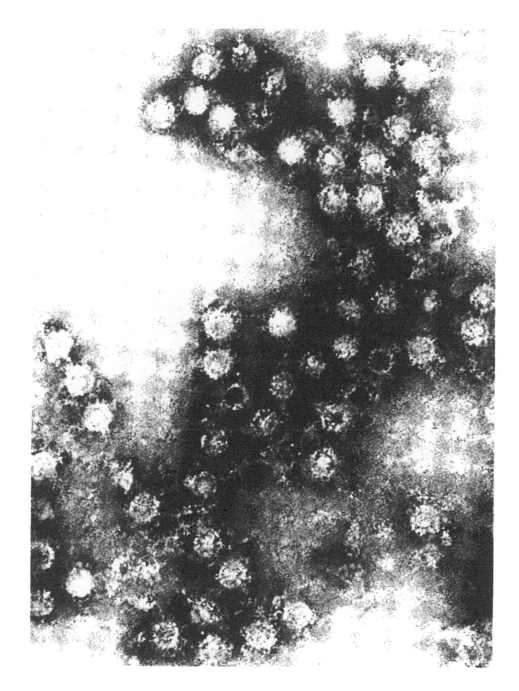

FIGURE 26.   Immune aggregate of coxsackie B virus from tissue culture fluid. UA stain × 298,620. (Courtesy of G. W. Gary, Jr., CDC.)

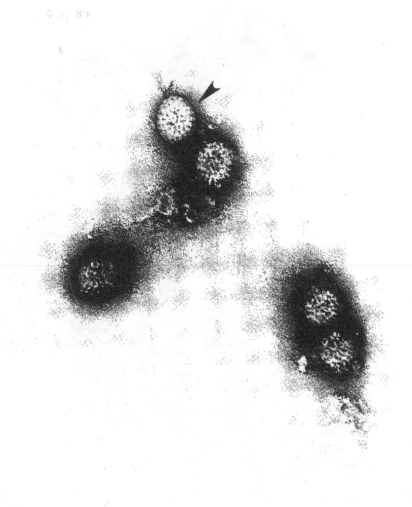

FIGURE 27.   IEM of rotavirus using monoclonal antibody to rotavirus group specific single-shelled particle surface antigen. Single-shelled particles are practically obscured by antibody, whereas there is no reaction with a double-shelled particle (arrow). UA stain × 175,793.

FIGURE 28.   Immune aggregate of enterovirus 70 from infected culture fluid. Particles are "buried" in antibody so that virus can be clearly seen only around the edge of the aggregate. UA stain × 156,260.

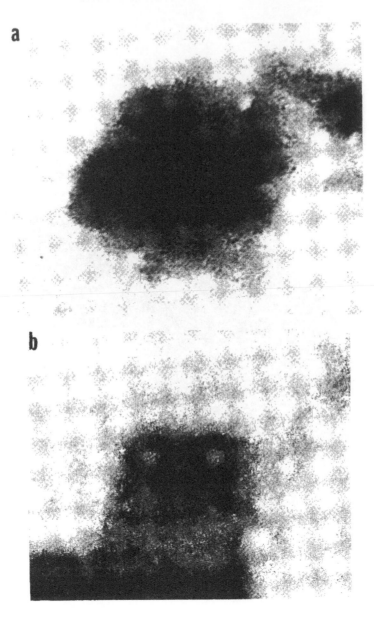

FIGURE 29.   IEM of enterovirus 70 from infected culture fluid. UA stain. (A) Immune aggregate with empty particles. × 195,325 (B) Immune aggregate of three particles. × 195,325.

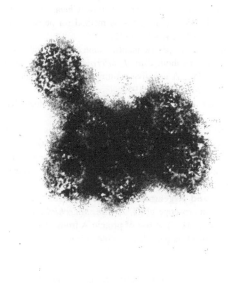

FIGURE 30.   IEM of human single shelled rotavirus particles.
Particle structure is obscured by a heavy layer of antibody. UA
stain × 237,680.

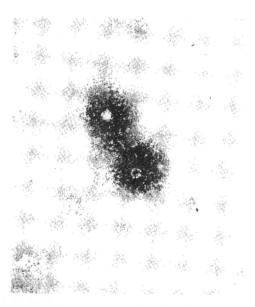

FIGURE 31.   Immune EM of parvovirus H-1 from rat liver
cells. Particles are very small and are almost obscured by antibody. UA stain × 156,260.

REFERENCES

1. **Anderson, T. F. and Stanley, W. M.,** A study by means of the electron microscope of the reaction between tobacco mosaic virus and its antiserum, *J. Biol. Chem.,* 139, 339, 1941.
2. **Anderson, N. and Doane, F. W.,** Agar diffusion method for negative staining of microbial suspensions in salt solutions, *Appl. Microbiol.,* 24, 495, 1972.
3. **Anderson, N. and Doane, F. W.,** Specific identification of enteroviruses by immuno-electron microscopy using a serum-in-agar diffusion method, *Can. J. Microbiol.,* 19, 585, 1973.
4. **Brenner, S. and Horne, R. W.,** A negative staining method for high resolution electron microscopy of viruses, *Biochim. Biophys. Acta,* 34, 103, 1959.
5. **Derrick, K. S.,** Quantitative assay for plant viruses using serologically specific electron microscopy, *Virology,* 56, 652, 1973.
6. **Lafferty, K. J. and Oertelis, S. J.,** The interaction between virus and antibody. III. Examination of virus-antibody complexes with the electron microscope, *Virology,* 21, 91, 1963.
7. **Luton, P.,** Rapid adenovirus typing by immunoelectron microscopy, *J. Clin. Pathol.,* 26, 914, 1973.
8. **Milne, R. G. and Luisoni, E.,** Rapid high-resolution immune electron microscopy of plant viruses, *Virology,* 68, 270, 1975.
9. **Nicolaieff, A., Obert, G., and Regenmortel, N. H. V.,** Detection of rotavirus by serological trapping on antibody coated electron microscope grids, *J. Clin. Microbiol.,* 12, 101, 1980.
10. **Shukla, D. D. and Gough, K. H.,** The use of protein A from *Staphylococcus aureus* in immune electron microscopy for detecting plant virus particles, *J. Gen. Virol.,* 45, 533, 1979.

REFERENCES FOR TABLE 3

1. **Almeida, J. D. and Goffe, A. P.,** Antibody to wart virus in human sera demonstrated by electron microscopy and precipitin tests, *Lancet,* 2, 1205, 1965.
2. **Bayer, M. E., Blumberg, B. S., and Werner, B.,** Particles associated with Australian antigen in sera of patients with leukemia, Down's syndrome and hepatitis, *Nature (London),* 218, 1057, 1968.
3. **Best, J. M., Banatvala, J. E., Almeida, J. D., and Waterson, A. P.,** Morphological characteristics of rubella virus, *Lancet,* 2, 237, 1967.
4. **Feinstone, S. M., Kapikian, A. Z., and Purcell, R. H.,** Hepatitis A: detection by immune electron microscopy of a virus-like antigen associated with acute illness, *Science,* 182, 1062, 1973.
5. **Kapikian, A. Z., Wyatt, R. G., Dolin, R., Thornhill, T. S., Kalica, A. R., and Chanock, R. M.,** Visualization by immune electron microscopy of a 27 nm particle associated with acute infectious nonbacterial gastroenteritis, *J. Virol.,* 10, 1075, 1972.
6. **McCormick, J. B., Sasso, D. R., Palmer, E. L., and Kiley, M. P.,** Morphological identification of the agent of Korean hemorrhagic fever (Hantaan virus) as a member of the Bunyaviridae, *Lancet,* 1, 765, 1982.

Chapter 4

MARKERS IN IEM: LIQUID PREPARATIONS

Two types of markers have been used to detect virus and virus-antibody interaction in liquid preparations. These are ferritin and colloidal gold. Ferritin is a protein enclosing an iron core. Colloidal gold is formed by reduction of chloroauric acid with sodium citrate. The electron density of these heavy metal markers results in an easily recognizable label appearing as small black dots in thin sections of cells. By negative stain EM, ferritin appears as small dark central structures and gold as opaque dots.

Ferritin conjugated to antibody combined with negative stain methods has been used to show the attachment of IgG on influenza virus and on hepatitis B core antigen. The method has also been used to detect rotavirus (Figure 32), adenovirus, and enterovirus. These viruses were detected using an indirect IEM method in which virus was complexed with antiserum then mixed with a species specific serum conjugated with ferritin. Heavy tagging of molecules of ferritin around individual particles facilitated virus detection. Ferritin can be conjugated directly to virus-specific serum or purchased commercially for use in the more sensitive indirect IEM technique. Conjugation of antibody with ferritin for the direct method has several drawbacks, including possible decrease of antibody titer during the covalent conjugation procedure.

Ferritin molecules are often seen in preparations of brain tissue centrifuged in density gradients and sometimes in serum. The molecules markedly resemble very small virus particles in morphology (Figure 33). Ferritin can be distinguished from virus particles by size, because the molecules are hollow and are approximately 15 nm in diameter which is smaller than any known virus.

Colloidal gold was initially used as a marker to locate cell surface antigen on thin sectioned cells. More recently, it has been adapted as a marker in liquid preparations. Grids coated with antiserum to tobamoviruses were used to trap particles on the grid. The grids were then floated on drops of a gold-protein A-antibody complex. This gold-labeled antibody decoration (GLAD) technique differentiated tobamoviruses by the amount of gold label attached to viruses.

Gold conjugated with serum containing antibody to hepatitis B antigen has been used as a marker to detect immune complexes of the antigen in concentrates of serum from HBsAg positive patients. Gold or gold-PA complexed with monoclonal antibody to the group specific antigen of human rotavirus can be used as a marker to identify the morphological structure associated with the antigen.

Binding of colloidal gold with antibody is electrostatic and does not require the damaging process of covalent linking used to conjugate ferritin with antibody. Particles can also be prepared in different sizes by varying the concentration of sodium citrate used to reduce chloroauric acid. Gold prepared by the sodium citrate reduction of chloroauric acid is electron dense as shown in Figure 34. After conjugation with protein, an electron lucent halo can be seen around particles (Figure 35). Gold-antibody complexes reactive with rotavirus and rotavirus subviral components are shown in Figures 36 and 37. Figure 38A is an immune aggregate of adenovirus hexon antigen. For comparison, a similar aggregate labeled with colloidal gold is shown in Figure 38B. Preparation of colloidal gold and conjugation of gold with protein has recently been reviewed by Roth.[10]

Another use of marker IEM is to detect antigen with no recognizable morphology by mixing the nondistinctive component with a marker such as a recognizable viral component. In this way, it is possible to form complexes of antigenically related structures with recognizable virus components. A low molecular weight subunit of rotavirus and a micellular form of hepatitis B surface antigen have been detected by this method. Tubular forms of polyomavirus and filamentous structures of rotavirus have been identified by a similar technique.

FIGURE 32.   IEM of rotavirus and rotavirus subcomponents labeled with ferritin. × 371,117.

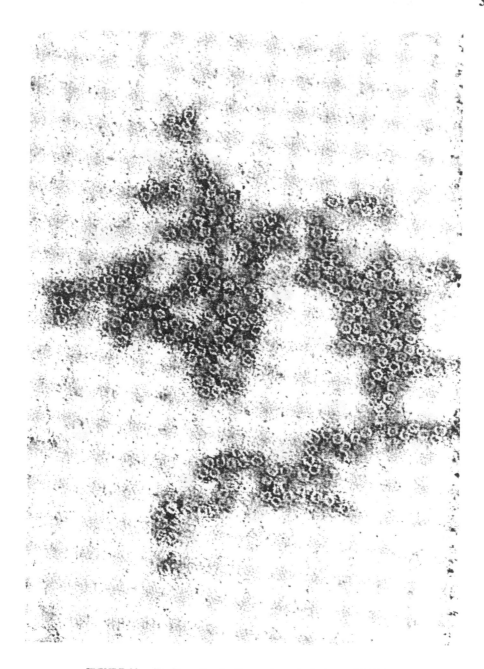

FIGURE 33.   Ferritin molecules from guinea pig brain. × 297,000.

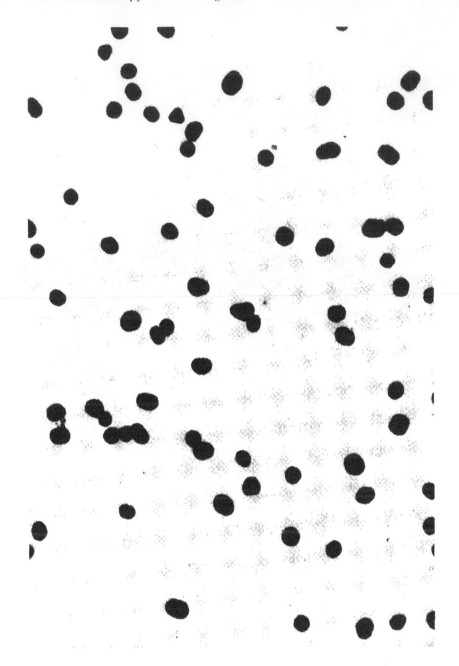

FIGURE 34.   Colloidal gold prepared by the reduction of chloroauric acid by sodium citrate
× 237,600.

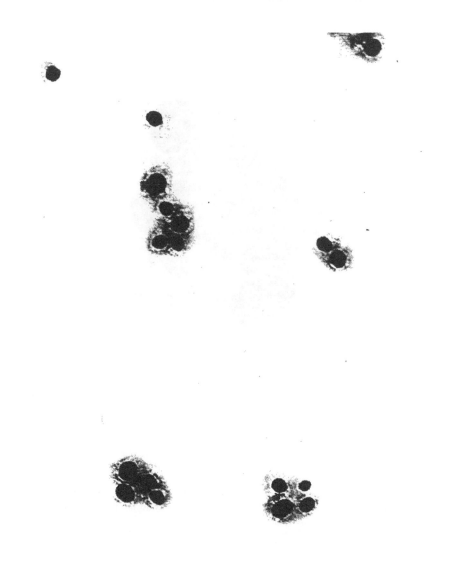

FIGURE 35. Colloidal gold complexed with protein. The protein appears as an electron lucent rim around the gold particles. × 195,325.

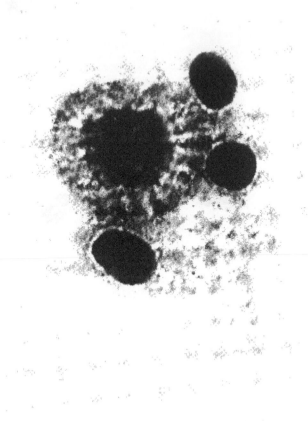

FIGURE 36.   Rotavirus single-shelled particle labeled with a conjugate of colloidal gold and antirotavirus monoclonal antibody to the group specific rotavirus antigen. UA stain × 612,900.

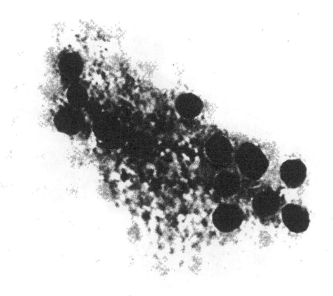

FIGURE 37. Subviral components of rotavirus labeled with a conjugate of colloidal gold and anti-rotavirus monoclonal antibody to the group specific rotavirus antigen. UA stain × 415,800.

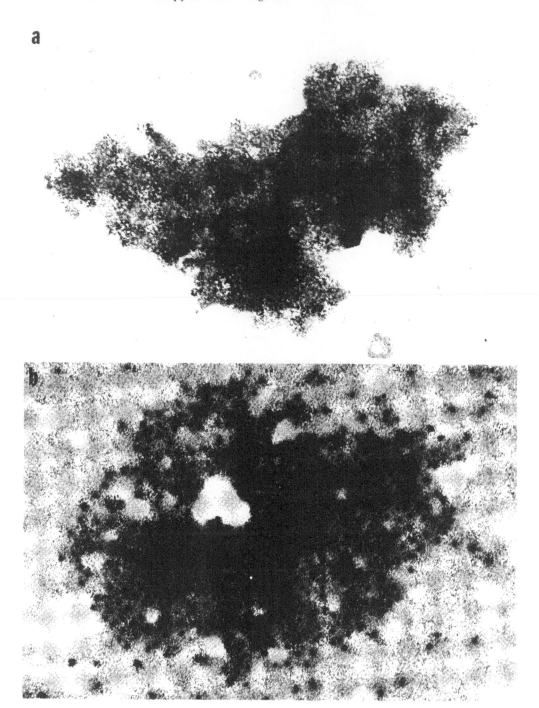

FIGURE 38. (A) Immune aggregate of adenovirus 2 hexon antigen. Hexons are aggregated by monoclonal antibody. UA stain × 102,240. (B) Immune aggregate of adenovirus 2 hexon antigen labeled with antihexon antibody conjugated with colloidal gold. UA stain × 156,260.

REFERENCES

1. **Almeida, J. D., Skelly, J., Howard, C. R., and Zuckerman, S.,** The use of markers in immune electron microscopy, *J. Virol. Methods,* 2, 169, 1981.
2. **Berthiaume, L., Alain, R., McLaughlin, B., Payment, P., and Trepanier, P.,** Rapid detection of human viruses by a simple indirect immune electron microscopy technique using ferritin-labelled antibodies, *J. Virol. Methods,* 2, 367, 1981.
3. **Faulk, W. P. and Taylor, G. M.,** An immunocolloid method for electron microscopy, *Immunochemistry,* 8, 1081, 1971.
4. **Huang, S. N. and Neurath, A. R.,** Immunohistologic demonstration of hepatitis B viral antigen in liver injury, *Lab. Invest.,* 40, 1, 1979.
5. **Holmes, I. H., Ruck, B. J., Bishop, R. F., and Davidson, G. P.,** Infantile enteritis viruses: morphogenesis and morphology, *J. Virol.,* 16, 937, 1975.
6. **Martin, M. L. and Palmer, E. L.,** Electron microscopic identification of rotavirus group antigen with gold-labelled monoclonal IgG, *Arch. Virol.,* 78, 279, 1983.
7. **Morgan, C., Refkind, R. A., Hsu, K. C., Holden, M., Segal, B. C., and Rose, H. M.,** Electron microscopic localization of intracellular viral antigen by the use of ferritin-conjugated antibody, *Virology,* 14, 292, 1961.
8. **Pares, R. D. and Whitcross, M. I.,** Gold labelled antibody decoration (GLAD) in the diagnosis of plant viruses by immuno-electron microscopy, *J. Immunol. Methods,* 51, 23, 1982.
9. **Patterson, S.,** Detection of antibody in virus-antibody complexes by immunoferritin labelling and subsequent negative staining, *J. Immunol. Methods,* 9, 115, 1975.
10. **Roth, J.,** The colloidal gold marker system for light and electron microscopic cytochemistry, in *Techniques in Immunocytochemistry,* Vol. 2, Bullock, G. R. and Petrusz, P., Eds., Academic Press, New York, 1983, 215.
11. **Stannard, L. M., Lennon, M., Hodgkiss, M., and Heidi, S.,** An electron microscopic demonstration of immune complexes of hepatitis B e-antigen using colloidal gold as a marker, *J. Med. Virol.,* 9, 165, 1982.

REFERENCES

Chapter 5

MARKERS IN THIN SECTION EM OF TISSUE AND CELLS

Immune EM of virus infected cells is used to detect virus or viral antigen on the surface of or within ultrathin sections of the cells. The type of marker used depends on the type, location, and stability of the antigen under study. Early techniques utilized the electron dense properties of ferritin conjugated with antiserum to localize antigens. It worked very well for labeling extracellular antigen prior to embedding, but the high molecular weight ferritin-antibody complexes would not adequately penetrate into cells unless permeabilization techniques which partially destroyed cell structure were used. The same problem occurred with antibody conjugated with other heavy metals such as mercury and uranium and with enzymes. This inherent problem has not been completely resolved, but postembedding labeling of thin sections to mark intracellular virus and antigen has been moderately successful in virology. The direct methods of both pre- and postembedding labeling for IEM have used antibody conjugated to metals. The indirect method uses a primary antiserum to label the antigen then anti-antibody conjugated to metals or enzymes to detect primary antigen-antibody complexes. A thin section of tissue culture cells infected with herpes simplex virus and indirectly labeled with gold prior to embedding is shown in Figure 39. Colloidal gold in complex with IgG or protein A has been introduced in the past few years for IEM. Figure 40 shows Hantaan virus in E-6 monkey kidney cells. The virus was labeled with specific antibody followed by colloidal gold complexed with protein A. Colloidal gold can be prepared in different sizes (5 to 20 nm) so that double labeling IEM is possible.

Antibody can also be conjugated with an enzyme such as horseradish peroxidase. This is a heme-containing glycoprotein of molecular weight 40.000. It functions as a marker by catalyzing the oxidation of 3,3' diaminobenzidine, a hydrogen donor, in the presence of hydrogen peroxide. It becomes electron dense by chelating diaminobenzidine with osmium tetroxide. The reaction can be intensified by performing it in phosphate buffer in the presence of cobalt chloride and nickel salts. Antibodies labeled with peroxidase can penetrate tissue sections more easily than those labeled with heavy metals. A thin section of canine kidney cells infected with influenza virus and labeled by the horseradish peroxidase method is shown in Figure 41. The biotin-avidin enzyme system can be used in the same way.

The plasma membrane of a cell is not easily permeable to antibody or antibody conjugates. To detect intracellular virus or viral components it is necessary to create holes in the membrane to allow antibody penetration. These are produced by a variety of methods such as detergent treatment, enzymatic action, and freeze-thaw. Each of these methods has disadvantages and a choice of permeabilizing agents has to be made by experimentation. One method has been developed using a saponin-aldehyde fixative. The procedure is described for IEM with peroxidase staining. It allows penetration of antibodies through cell membranes, provides good cell preservation, and does not destroy viral antigenicity to a great extent. It consists of pretreating cells with 0.05% saponin (a soap bark detergent), 0.0125 to 0.05% glutaraldehyde, and 1% paraformaldehyde for 5 min at 4°C, and then postfixing with the same fixative but without saponin for 45 min at 4°C. Bohn was able to define the intracellular development of Shope fibroma, a poxvirus, and the appearance of viral antigen at the cell membrane using the saponin-aldehyde fixative and immunoperoxidase staining. We used the method with minor modifications to detect antigen along the plasma membrane of T lymphocytes infected with human immunodeficiency virus (AIDS) and to describe development of the viral nucleoid (Figures 42 and 43). This retrovirus has a glycoprotein surface which is apparently not destroyed by the procedure. Most other enveloped viruses also have glycoprotein surfaces and these glycoproteins are altered by Nonidet P40 and other nonionic detergents commonly used in virology. Permeabilization of cells with saponin following mild aldehyde fixation has also been used successfully to localize intracellular rotavirus antigens using colloidal gold as an electron dense marker. Thus, saponin appears to be a useful detergent for IEM of cells infected with some viruses.

Immune EM has also been used to detect antigen in postembedded tissue. Antigen is exposed by "etching" of epoxy resins with hydrogen peroxide or alcohols and of methacrylate with benzene or xylene. It is difficult to maintain both ultrastructural preservation and viral antigenic activity during fixation of cells for embedment. The best fixation for a given antigen must be determined for each set of circumstances. Numerous fixative formulas have been used. The most successful appear to be some combinations of glutaraldehyde-paraformaldehyde with glutaraldehyde concentration kept between 0.5 and 2.0%. Cells are fixed, embedded, "etched" with hydrogen peroxide, and exposed to antibody conjugated to heavy metals or enzymes. Colloidal gold and horseradish peroxidase are currently the most widely used markers for postembedding labeling. Copper reacts with osmium tetroxide and hydrogen peroxide so nickel or gold grids need to be used in postembedding IEM procedures.

There are a number of choices available for detecting antigen in either pre- or postembedded tissue. Some of these are

1. The direct antibody method where antibody conjugated with marker is added directly to cells.
2. The indirect method where antivirus antibody is reacted with the tissue followed by anti-antibody conjugated with marker.
3. A "bridge" method. Unconjugated antibody is reacted with cells. Then antibody against the initial antibody is added. An antibody produced in the same species as the first antibody, but conjugated with marker, is reacted with any unoccupied binding site of the anti-antibody.

Other amplifications are possible and procedures using protein A and immunoglobulin conjugated markers have been used successfully in virology. *In situ* hybridization using labeled DNA and RNA probes has also been used to localize viral nucleic acid within thin sectioned cells. It is expected that more widespread use of various marker techniques will greatly improve localization of viral antigens and replicative sites.

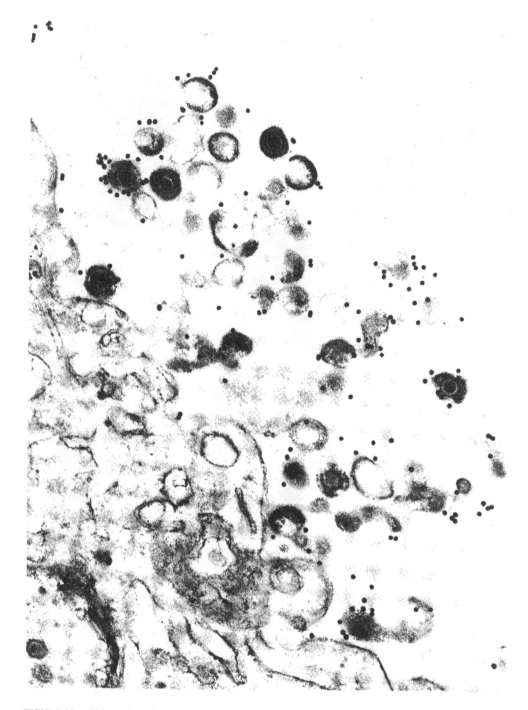

FIGURE 39.   Thin section of human lung fibroblast infected with herpes simplex virus. Virus was labeled with specific antibody followed by anti-antibody labeled with colloidal gold. UA stain × 155,000.

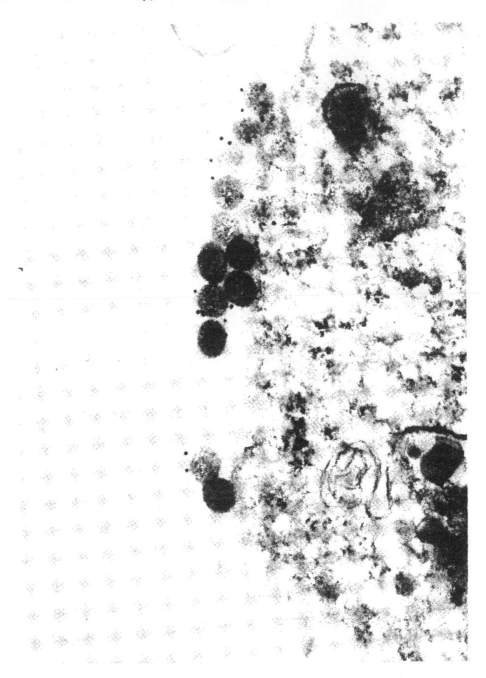

FIGURE 40.   IEM of Hantaan virus infected E-6 VERO cells. Anti-Hantaan virus antibody was added to infected cells followed by colloidal gold conjugated with protein A. × 83,700. (Courtesy of John White, USAMRID.)

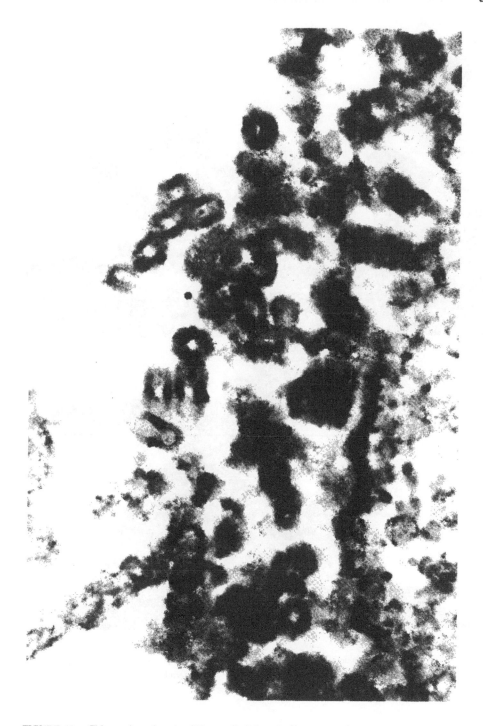

FIGURE 41.   Thin section of canine kidney cells infected with type A influenza virus. Virus was labeled by the horseradish peroxidase method. Reactivity is indicated by build-up of electron dense reaction product. × 83,700.

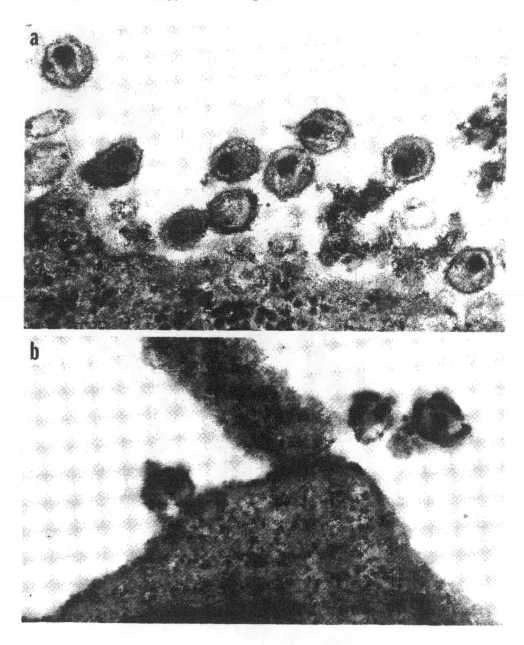

FIGURE 42.   Human immunodeficiency virus (AIDS) in an extracellular space (A) unlabeled and (B) labeled with horseradish peroxidase. The electron dense nucleoid is still discernible. × 95,480.

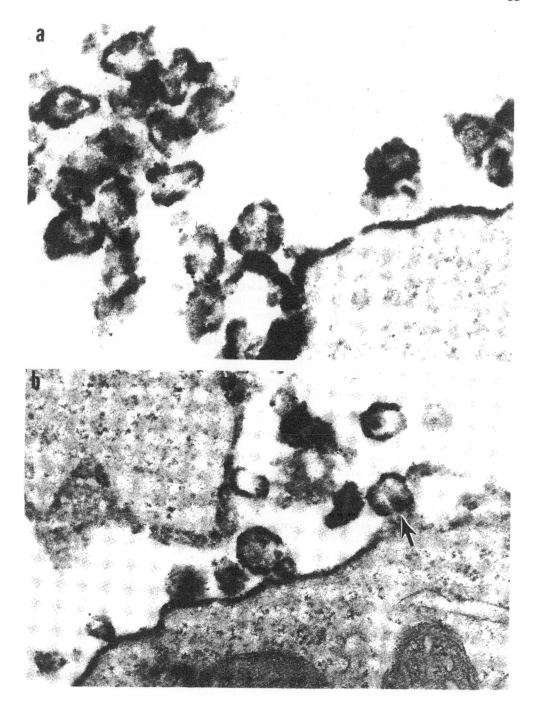

FIGURE 43.   IEM with peroxidase labeling of human immunodeficiency virus (AIDS) infected T4 lymphocytes.
(A) The plasma membrane is heavily labeled as is the envelope of extracellular particles. × 87,420. (B) The
arrow in this figure points to a newly forming particle which is labeled with peroxidase. The nucleoid is formed
when the particle buds from the plasma membrane. × 87,420.

REFERENCES

1. **Adams, J. C.,** Heavy metal intensification of DAB based HRP reaction product, *J. Histochem. Cytochem.,* 29, 775, 1981.
2. **Bendayan, M. and Zollinger, M.,** Ultrastructural localization of antigenic sites on osmium-fixed tissues applying the protein A-gold technique, *J. Histochem. Cytochem.,* 31, 101, 1983.
3. **Bohn, W.,** A fixation method for improved antibody penetration in electron microscopical immunoperoxidose studies, *J. Histochem.,* 26, 293, 1978.
4. **Bohn, W.,** Electron microscopic immunoperoxidase studies on the accumulation of virus antigen in cells infected with Shope fibroma virus, *J. Gen. Virol.,* 46, 439, 1980.
5. **Broker, T. R., Angerer, L. M., Yen, P. H., Hershey, N. D., and Davidson, N.,** Electron microscopic visualization of tRNA genes with ferritin-avidin labels, *Nucleic Acids Res.,* 5, 363, 1978.
6. **Gelderblom, H., Kocks, C., L'age-Stehr, J., and Reupke, H.,** Comparative immunoelectron microscopy with monoclonal antibodies on yellow fever virus-infected cells: pre-embedding labelling versus immunocryoultramicrotomy, *J. Virol. Methods,* 10, 225, 1985.
7. **Geoghegan, W. D. and Ackerman, G. A.,** Adsorption of horseradish peroxidase, ovomucoid and anti-immunoglobulin to colloidal gold for the indirect detection of concanavalin A, wheat germ agglutinin and goat anti-human immunglobulin G on the cell surface at the electron microscopic level: a new method, theory and application, *J. Histochem. Cytochem.,* 25, 1187, 1977.
8. **Geuze, H. J., Slot, J. W., Vanderley, P., and Scheffer, R. C. T.,** Use of colloidal gold particles in double-labeling immunoelectron microscopy of ultrathin frozen tissue section, *J. Cell. Biol.,* 89, 653, 1981.
9. **Hayat, M. A., Ed.,** *Electron Microscopy of Enzymes: Principles and Methods,* Van Nostrand Reinhold Co., New York, 1975.
10. **Hutchison, N. J., Langer-Safer, P. R., Ward, D. C., and Hamkalo, B. A.,** *In situ* hybridization at the electron microscope level: hybrid detection by autoradiography and colloidal gold, *J. Cell Biol.,* 95, 609, 1982.
11. **Morgan, C., Hsu, K. C., Refkin, O. R. A., Knox, A. W., and Rose, H. M.,** The application of ferritin-conjugated antibody to electron microscopic studies of influenza virus infected cells, *J. Exp. Med.,* 111, 833, 1981.
12. **Murata, F., Suganuma, T., Tsuyama, S., Ishida, K., and Funasako, S.,** Glycoconjugate cytochemistry of the rat small intestine using *Helix pomatia* agglutinin and colloidal gold conjugators, *J. Electron Microsc.,* 35, 29, 1986.
13. **Nakane, P. K.,** Immunoelectron microscopy, *Methods Cancer Res.,* 20, 183, 1982.
14. **Narayanswami, S. and Hamkalo, B.,** Electron microscopic *in situ* hybridization using biotinylated probes, *Focus,* 8, 3, 1986.
15. **Pickel, V. M.,** Single and dual localization of neuronal antigens using enzymatic, gold, and autoradiographic markers, *EMSA Bull.,* 16, 61, 1986.
16. **Roth, J., Bendayan, M., and Orci, L.,** Ultrastructural localization of intracellular antigens by the use of protein A-gold complex, *J. Histochem. Cytochem.,* 26, 1074, 1978.

Chapter 6

GASTROENTERITIS VIRUSES

The Norwalk agent and rotavirus are the major causes of nonbacterial gastroenteritis in humans. Both viruses were identified by transmission EM in the early 1970s.

The Norwalk agent, a virus 27 nm in diameter, has not been assigned to a taxonomic group. It was initially detected by IEM in stools of patients with nonbacterial gastroenteritis in Norwalk, Ohio. The virus has been seen only by negative stain IEM, and antibody attached to the surface of the virus has obscured its ultrastructure. The Norwalk agent is responsible for large scale, explosive epidemics of gastroenteritis in adults and children over 5 years of age. It is usually food or water borne with some secondary person-to-person spread. Other viruses which morphologically resemble the Norwalk agent, but are antigenically different, have also been associated with large-scale epidemics. These viruses are difficult to detect by EM because particles are shed in small quantities and none have been cultivated in vitro. IEM remains the touchstone for the identification of these viruses. Viruses associated with nonbacterial gastroenteritis are listed in Table 4.

Rotavirus primarily affects children under 5 years of age. The virus was initially detected in duodenal biopsy tissue by thin section EM and, soon after, by direct negative stain EM of stools. The virus occurs as either a double or single shelled form. Double-shelled particles have a distinct wheel-like (rota meaning wheel) structure which is easily recognizable by EM. Particles without the outer shell (single shelled) have a surface structure with large ring-like subunits which are thought to be arranged in a $T = 7$ skew arrangement. Large numbers of particles are shed in feces so that rapid diagnosis of rotavirus gastroenteritis is feasible by EM. Stool suspensions, usually 20%, are remarkably free of large amounts of electron dense background debris and are among the easiest body fluids or excretions to examine by EM. Detection of rotavirus by the pseudoreplica negative stain technique can be accomplished in about 10 min if the specimen contains at least 10^6 particles/mℓ. During acute rotavirus diarrhea, particle counts of 10^{10}/mℓ are not uncommon.

In addition to rotaviruses and the Norwalk and Norwalk-like viruses there are numerous other viruses which are putative agents of nonbacterial gastroenteritis. These include adenoviruses, astrovirus, calicivirus, coronavirus, reovirus, and other morphologically less well defined small, round viruses. Bacteriophage are also seen in stool specimens, and those that are small and round can easily be mistaken for possible gastroenteritis agents. Examples of viruses found in stools of patients with nonbacterial gastroenteritis are shown in Figures 44 to 47.

The involvement of adenoviruses in viral gastroenteritis was established after the finding of large numbers of particles morphologically resembling adenovirus in stools of persons ill with nonbacterial gastroenteritis. In early studies it was found that these adenoviruses were not cultivable by the techniques commonly used for respiratory adenoviruses; thus these viruses were labeled "noncultivable" and subsequently "fastidious" after it was determined that the viruses are cultivable, but only under special circumstances. Fastidious adenoviruses of stool origin that were propagated under these conditions were found to be different from the 39 previously known species by neutralization tests. There are now at least two new provisional types associated with viral gastroenteritis; these are Ad40 and Ad41.

Astroviruses are 28 to 30 nm in diameter and have a characteristic 5 or 6 pointed star-shaped surface structure. The virus is thought to cause gastroenteritis in infants, children, and adults. Astrovirus has not been cultivated in vitro but is excreted in large numbers, and when present in stools is easily recognizable by a star-shaped morphology. However, this configuration may sometimes be obscured by antibody when examining stools by IEM.

Caliciviruses have a surface composed of cup-like depressions which are characteristic for this virus. These depressions can also be seen in outline around the periphery of the

Table 4
VIRUSES DETECTED BY IEM
IN STOOLS OF PATIENTS
WITH NONBACTERIAL
GASTROENTERITIS

Group I	
Norwalk-like agents	**Ref.**
Norwalk	7
Harlow	2
Minireovirus	9
Montgomery County	13
Hawaii	13
Taunton	4
Otofuke	12
Sapporo	8
Marin County	10
Snow Mountain	6

Group II	
Small round unstructured viruses	**Ref.**
Wollan	11
Ditchling	1
Cockle	3
Parramatta	5

particles. Caliciviruses have been associated with outbreaks of gastroenteritis, primarily in infants and children.

Coronaviruses have been detected in stools of persons with gastroenteritis as well as asymptomatic children and adults. In some cases pieces of presumable cell debris have been falsely identified as coronavirus (pseudocoronavirus); therefore, the relationships between coronaviruses and gastroenteritis in humans remains to be clarified.

The role of other viruses detected in feces by EM such as "minirotavirus" and "mini-reovirus" has not been adequately evaluated. More definitive studies await in vitro cultivation of the numerous "viruses" now associated with nonbacterial gastroenteritis. Rotavirus and the enteric adenoviruses are the only gastroenteritis viruses of humans which have been cultivated in cell culture. Meanwhile negative stain EM and IEM remain the most useful methods for detection of gastroenteritis viruses other than rotavirus. Commercial enzyme-linked immunosorbent assay (ELISA) kits are available for detection of rotavirus in stools.

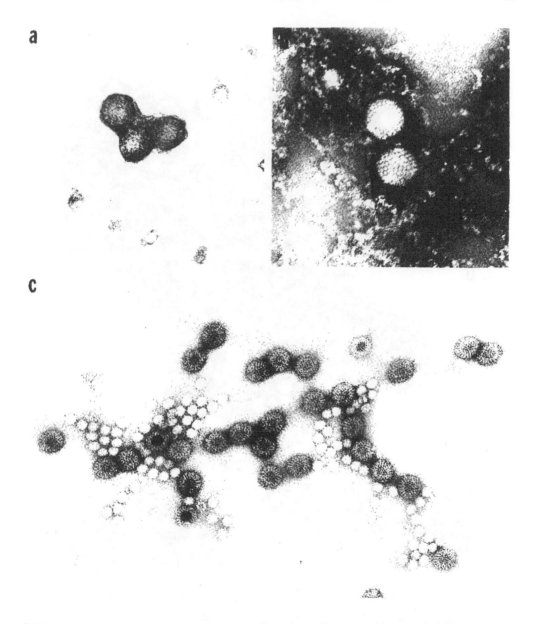

FIGURE 44.   Negatively stained (UA) viruses found in stools of patients with gastroenteritis. (A) Reovirus, × 136,728; (B) Adenovirus, × 155,000; and (C) Rotavirus and small round 32 nm particle, × 89,460.

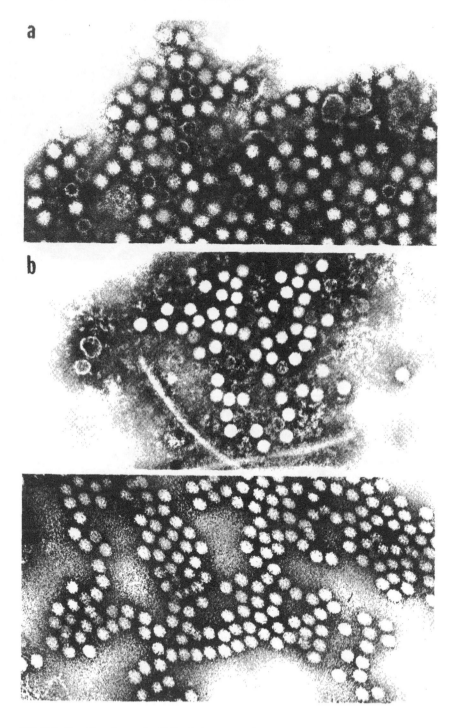

FIGURE 45.   Negative stain (UA) of virus in stools from patients with gastroenteritis. (A) Calicivirus, × 117,195; (B and C) small round structured viruses 27 to 29 nm in diameter, (B) × 97,663, (C) × 115,020.

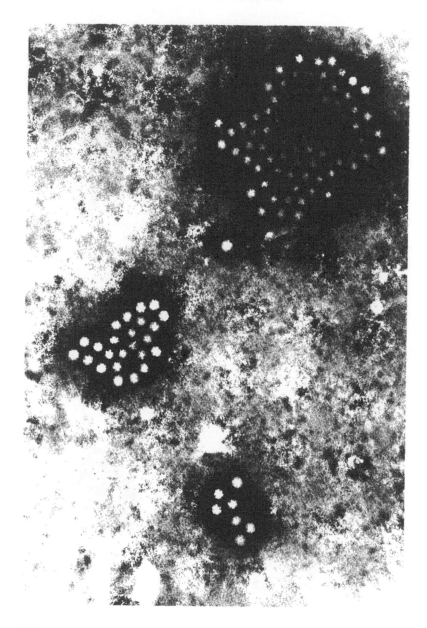

FIGURE 46. Astrovirus, negatively stained with PTA, in stool of a young girl who was ill with gastroenteritis. × 89,460. (Courtesy of Fred Williams, USEPA.)

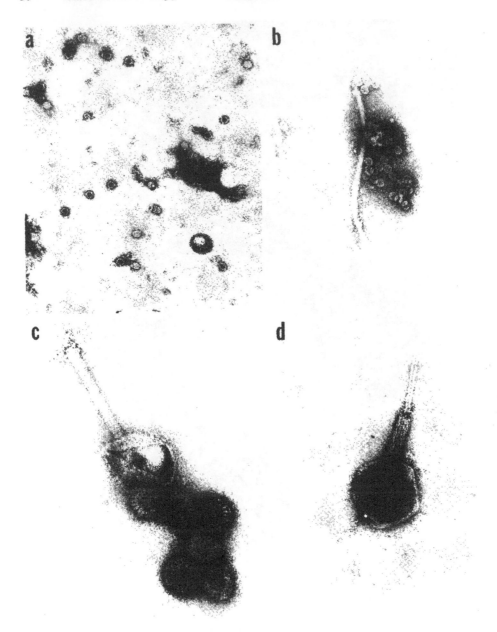

FIGURE 47.   Negatively stained (UA) bacterial viruses found in stools of patients with gastroenteritis. (A) Bacteriophage φX174, × 93,663; (B) bacteriophage MS2, × 78,130; (C) T-even bacteriophage and rotavirus, × 97,663; and (D) T-even bacteriophage, × 97,663.

REFERENCES

1. **Caul, E. O. and Appleton, H.,** The electron microscopical and physical characteristics of small round human fecal viruses: an interim scheme for classification, *J. Med. Virol.,* 9, 257, 1982.
2. **Caul, E. D., Paver, W. K., and Clarke, S. K. R.,** Coronavirus particles in faeces from patients with gastroenteritis, *Lancet,* 1, 1192, 1975.
3. **Flewett, T. H., Bryden, A. S., and Davies, H.,** Virus particles in gastroenteritis, *Lancet,* 2, 1497, 1973.
4. **Gary, G. W., Jr., Hierholzer, J. C., and Black, R. E.,** Characteristics of noncultivable adenoviruses associated with diarrhea in infants: a new subgroup of adenoviruses, *J. Clin. Microbiol.,* 10, 96, 1979.
5. **Graham, F. L., Smiley, J., Russell, W. C., and Nairn, R.,** Characteristics of a human cell line transformed by DNA from human adenovirus type 5, *J. Gen. Virol.,* 36, 59, 1977.
6. **Kapikian, A. Z., Wyatt, R. G., Dolin, R., Thornhill, T. S., Kalica, A. R., and Chanock, R. M.,** Visualization by immune electron microscopy of a 27-nm particle associated with acute nonbacterial gastroenteritis, *J. Virol.,* 10, 1075, 1972.
7. **Kidd, A. H., Banatvala, J. E., and de Jong, J. C.,** Antibodies to fastidious faecal adenoviruses (species 40 and 41) in sera from children, *J. Med. Virol.,* 11, 333, 1983.
8. **Madeley, C. R. and Cosgrove, B. P.,** Viruses in infantile gastroenteritis, *Lancet,* 2, 124, 1975.
9. **Madeley, C. R. and Cosgrove, P. B.,** Calicivirus in man, *Lancet,* 1, 199, 1976.
10. **de Jong, J. C., Wigand, R., Kidd, A. H., Wadell, G., Kapsenberg, J. G., Muzerie, C. J., Wermenbol, A. G., and Firtzlaff, R. G.,** Candidate adenoviruses 40 and 41: fastidious adenoviruses from human infant stool, *J. Med. Virol.,* 11, 215, 1983.
11. **Takiff, H. E., Strauss, S. E., and Garon, C. F.,** Propagation and in vitro studies of previously noncultivable entero adenoviruses in Graham 293 cells, *Lancet,* 2, 832, 1981.
12. **Uhnoo, I., Wadell, G., Svensson, L., and Johansson, M. E.,** Two new serotypes of enteric adenovirus causing infantile gastroenteritis, *Dev. Biol. Stand.,* 53, 311, 1983.

REFERENCES FOR TABLE 4

1. **Appleton, H., Buckley, M., Thom, B. T., Cotton, J. L., and Henderson, S.,** Virus-like particles in winter vomiting disease, *Lancet,* 1, 409, 1977.
2. **Appleton, H. and Higgins, A.,** Viruses and gastroenteritis in infants, *Lancet,* 1, 1297, 1975.
3. **Appleton, H. and Pereira, M. S.,** A possible virus etiology in outbreaks of food-poisoning from cockles, *Lancet,* 1, 780, 1977.
4. **Caul, E. A., Ashley, C. R., and Pether, J. V. S.,** "Norwalk-like" particles in epidemic gastroenteritis in the UK, *Lancet,* 2, 1292, 1979.
5. **Christopher, P. J., Grohmann, G. S., Millsom, R. H., and Murphy, A. M.,** Parvovirus gastroenteritis — a new entity for Australia, *Med. J. Aust.,* 1, 121, 1978.
6. **Dolin, R., Reichman, R. C., Roessner, K. D., Tralka, T. S., Schooley, R. T., Gary, W., and Morens, D.,** Detection by immune electron microscopy of the Snow Mountain agent of acute viral gastroenteritis, *J. Infect. Dis.,* 146, 184, 1982.
7. **Kapikian, A. Z., Wyatt, R. G., Dolin, R., Thornhill, T. S., Kalica, A. R., and Chanock, R. M.,** Visualization by immune electron microscopy of a 27-nm particle associated with acute infectious nonbacterial gastroenteritis, *J. Virol.,* 10, 1075, 1972.
8. **Kogasaka, R., Sakuma, Y., Chiba, S., Akihara, M., Horino, K., and Nakao, T.,** Small round virus-like particles associated with acute gastroenteritis in Japanese children, *J. Med. Virol.,* 5, 151, 1980.
9. **Middleton, P. J., Szymanski, T., and Petric, M.,** Viruses associated with acute gastroenteritis in young children, *Am. J. Dis. Child.,* 131, 733, 1977.
10. **Oshiro, L. S., Haley, C. E., and Roberts, R. R.,** A 27 nm virus isolated during an outbreak of acute infectious nonbacterial gastroenteritis in a convalescent hospital: a possible new sterotype, *J. Infect. Dis.,* 143, 791, 1981.
11. **Paver, W. K., Caul, E. O., and Clarke, S. K. R.,** Comparison of a 22 nm virus from human faeces with animal parvoviruses, *J. Gen. Virol.,* 22, 447, 1974.
12. **Tanigucki, K., Urasawa, S., and Urasawa, T.,** Virus-like particle, 35- to 40-nm associated with institutional outbreak of acute gastroenteritis in adults, *J. Clin. Microbiol.,* 10, 730, 1979.
13. **Thornhill, T. S., Wyatt, R. G., Kalica, A. R., Dolin, R., Chanock, R. M., and Kapikian, A. Z.,** Detection by immune electron microscopy of 26- to 27-nm virus-like particles associated with two family outbreaks of gastroenteritis, *J. Infect. Dis.,* 135, 20, 1977.

Chapter 7

PICORNAVIRIDAE

Morphologically, picornaviruses are small (22 to 30 nm) icosahedrons with cubic capsid symmetry. The capsid is thought to be formed by 60 subunits which enclose a genome of ssRNA. There is no envelope or surface projections and the surface of these viruses is almost featureless. All members of the Picornaviridae have identical morphology. Negatively stained preparations show icosahedral particles of relatively uniform size and shape (Figure 48).

The genera of Picornaviridae which infect humans are *Enterovirus* and *Rhinovirus*. Species of *Enterovirus* include polioviruses, ECHO viruses, coxsackieviruses, and more than 70 human enteroviruses. Rhinoviruses are important causes of respiratory infections (common colds) and consist of over 100 serotypes. Hepatitis A virus is also thought to be a picornavirus.

Direct EM can be used to group viruses of the Picornaviridae but serotyping must be done by other methods because the viruses have identical morphology. IEM has been successfully applied to serotyping of some enteroviruses. The technique is useful for identifying hepatitis A virus isolates from stools and acute hemorrhagic conjunctivitis virus (enterovirus 70) from tears and eye swabs after passage in tissue culture.

Picornaviruses contain a single strand of positive-strand RNA which acts as a messenger for the synthesis of viral proteins and an RNA replicase. To initiate infection, these positive stranded viruses attach to receptors of susceptible cells. Penetration occurs and the viral RNA is rapidly uncoated. The RNA is translated monocistronically into a large polypeptide which is then processed into specific viral proteins. Replication occurs in the cytoplasm and the final stage of assembly is the combination of RNA and a shell of viral protein (the procapsid). Virus is released through vacuoles or in a burst by lysis of infected cells. A stylized drawing representing replication of picornaviruses is shown in Figure 49. Similar, but not exact, replicative steps are involved with other viruses which have positive stranded ssRNA genomes. The positive stranded RNA-containing viruses include the Picornaviridae, Togaviridae, Flaviviridae, Coronaviridae, and Caliciviridae.

Marked changes appear in the cytoplasm of picornavirus-infected cells where large numbers of membrane-bound pieces of cytoplasm accumulate in the central region of the cell. The nucleus is displaced and eosinophilic bodies develop in the cytoplasm. These are thought to be the site of assembly of viral subunits. Virus is first seen within and between cytoplasmic bodies and large crystals of virus may form as seen in Figures 50 and 51. Picornaviruses may orient in columns supported by a filamentous lattice such as the arrangement in Figure 52.

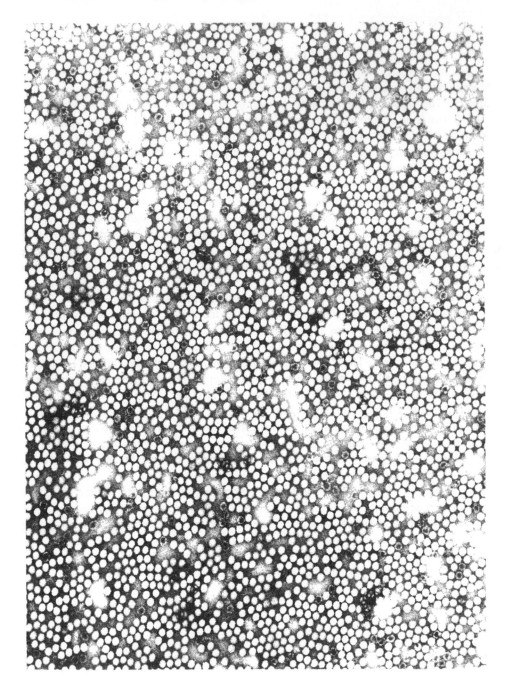

FIGURE 48.   Poliovirus negatively stained with UA. Both full (light) and empty (dark) particles have a hexagonal shape. Surface structure is not resolved. Particles are generally about the same in diameter. × 70,500.

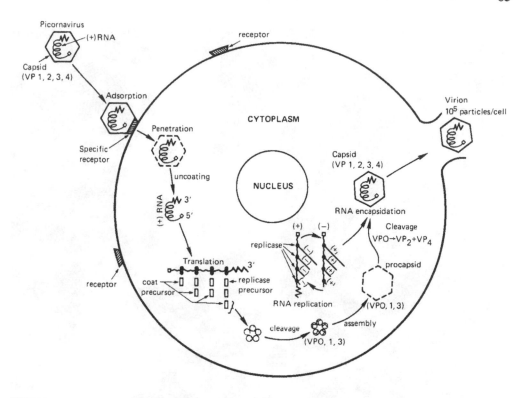

FIGURE 49.   Schematic showing overall events occurring during the replication of a picornavirus. Uncoated RNA acts as messenger to direct the synthesis of viral progeny. Virus is usually released by lysis of cells.

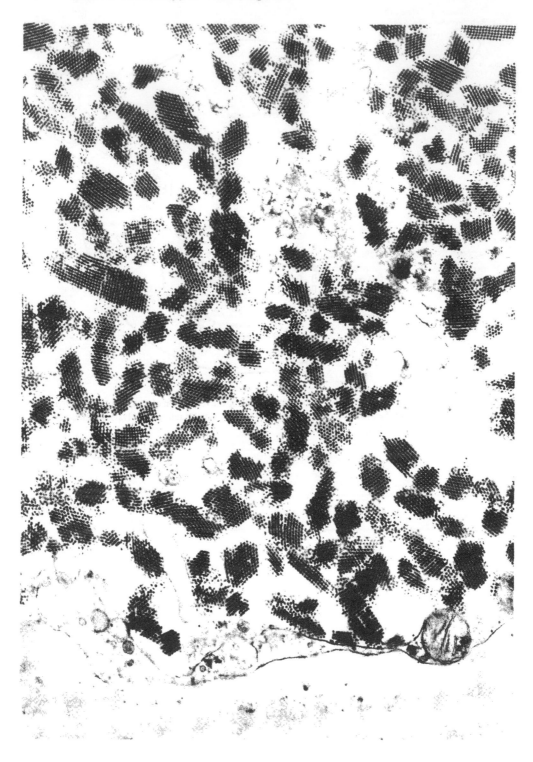

FIGURE 50   Nodamura virus, an arthropod transmissible picornavirus in mouse hind limb muscle. Virus crystaloids have entirely replaced the cytoplasm of a muscle cell. × 19,000. (Courtesy of Frederick Murphy, CDC.)

FIGURE 51. Hexagonal packing of complete and empty coxsackie A4 virus in mouse hind limb muscle. × 58,750. (Courtesy of Alyne Harrison, CDC.)

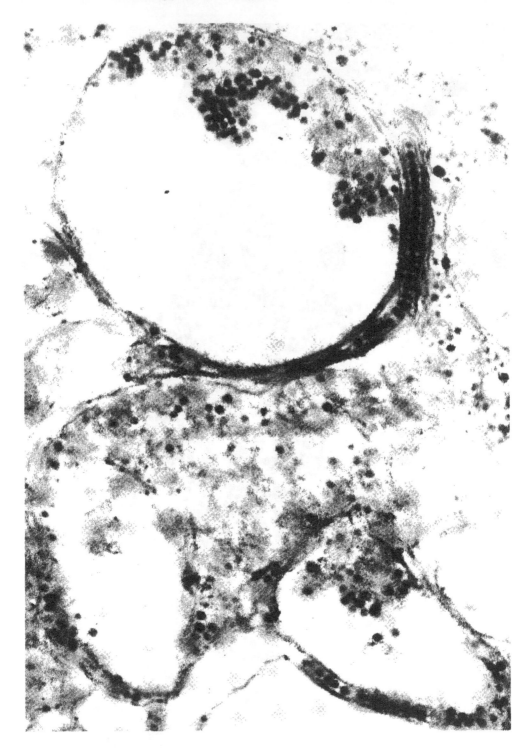

FIGURE 52.   Coxsackie A4 virus in linear fashion between membranes of mouse hind limb muscles. The field also shows particles in a random array within a finely granular matrix. × 66,900. (Courtesy of Alyne Harrison, CDC.)

REFERENCES

1. **Almeida, J. D., Waterson, A. P., Prydie, J., and Fletcher, E. W. L.,** The structure of picornavirus and its relevance to cubic viruses in general, *Arch. Gesamte Virusforsch.,* 25, 105, 1968.
2. **Baltimore, D.,** The replication of picornaviruses, in *Biochemistry of Viruses,* Levy, H. B., Ed., Marcel Dekker, New York, 1969, 101.
3. **Levintow, L.,** Reproduction of picornaviruses, in *Comprehensive Virology,* Vol. 2, Fraenkel-Conrat, H. and Wagner, R. R., Eds., Plenum Press, New York, 1974, 109.
4. **Mayor, H.,** Picornavirus symmetry, *Virology,* 22, 156, 1964.
5. **McGregor, S., Hall, L., and Rueckert, R. R.,** Evidence for the existence of protomers in the assembly of encephalomyocarditis virus, *J. Virol.,* 15, 1107, 1975.
6. **Rueckert, R. R.,** Picornaviral architecture, in *Comparative Virology,* Maramorosch, K. and Kurstak, E., Eds., Academic Press, New York, 1971, 255.
7. **Rueckert, R. R.,** On the structure and morphogenesis of picornaviruses, in *Comprehensive Virology,* Vol. 6, Fraenkel-Conrat, H. and Wagner, R. R., Eds., Plenum Press, New York, 1976, 131.
8. **Zwillenberg, L. O. and Burki, F.,** On the capsid structure of some small feline and bovine RNA viruses, *Arch. Gesamte Virusforsch.,* 19, 373, 1966.

REFERENCES

Chapter 8

CALICIVIRIDAE

The caliciviruses are ssRNA viruses which replicate in the cytoplasm of infected cells. By negative stain EM particles are round and measure 35 to 39 nm in diameter. The surface of the particle has 32 cup shaped depressions arranged in icosahedral symmetry. In certain orientations, these depressions cause six electron lucent areas to be visible over the particle periphery giving a ''Star of David'' appearance which is characteristic of Caliciviridae.

Viruses with calicivirus morphology as shown in Figure 53 have been found in association with gastroenteritis in swine, calves, and humans. Limited IEM studies indicate a relationship between calicivirus-like particles and gastroenteritis. The type species of the Caliciviridae is vesicular exanthema of swine. There are several types of feline calicivirus, and the virus is also found in San Miguel sea lions. Caliciviruses are positive stranded RNA viruses with more than one species of RNA. The RNA is infectious and the virion appears to be composed of only one major and one minor protein. Cytoplasmic granular accumulations associated with smooth membrane bound vesicles are thought to be the site of synthesis of calicivirus precursors. Virus accumulates in clusters in crystalline arrays and may be associated with cisternae. The Golgi apparatus may also be altered. Viroplasm accumulates in association with smooth membrane-bound vesicles and virus may be released by cell lysis or through vacuoles. However, the exact mechanism for replication and release of viruses of the Caliciviridae has not been completely defined. The caliciviruses associated with gastroenteritis in humans have not been cultivated in vitro.

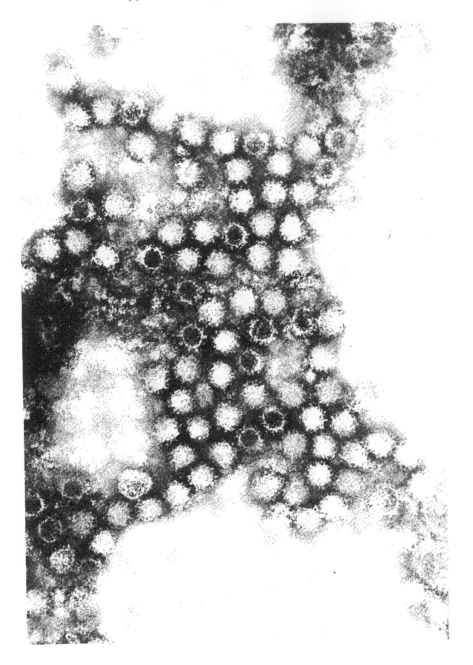

FIGURE 53.   Calicivirus from infected cat tongue cells negatively stained with UA. The surface of calicivirus has cup-like depressions.   × 195,325.

REFERENCES

1. **Schaffer, F. L.,** Caliciviruses, in *Comprehensive Virology,* Vol. 14, Frankel-Conrat, H. and Wagner, R. R., Eds., Plenum Press, New York, 1979.
2. **Zee, Y. C., Hackett, A. J., and Madin, S. H.,** Electron microscopic studies on vesicular exanthema of swine virus: intracytoplasmic viral crystal formation in cultured pig kidney cells, *Am. J. Vet. Res.,* 29, 1025, 1968.
3. **Zee, Y. C., Hackett, A. J., and Talens, L. T.,** Electron microscopic studies on the vesicular exanthema of swine virus. Morphogenesis of VESV type H_{54} in pig kidney cells, *Virology,* 34, 596, 1968.
4. **Zwillenberg, L. O. and Burki, F.,** On the capsid structure of some small feline and bovine RNA viruses, *Arch. Gesamte Virusforsch.,* 19, 373, 1966.

Chapter 9

REOVIRIDAE

The genera of Reoviridae which have species infecting humans include *Reovirus, Orbivirus,* and *Rotavirus.* These viruses have genomes consisting of dsRNA which is segmented into ten or more segments.

Most mammalian virus families consist of genera of viruses with identical morphology (*Enterovirus,* etc.). The Reoviridae, however, is an exception to this general rule because each of the three genera infecting vertebrates has a distinctive morphology. Viruses of each genus can be differentiated from viruses of other genera by direct negative stain EM.

Reovirus type 1 is about 75 nm in diameter and has a double-shelled capsid (Figure 54). When the outer capsid layer is digested away with proteolytic enzymes a core is released. The core has 12 spikes with 5-fold symmetry arranged icosahedrally. The core is 52 nm in diameter and encloses the dsRNA genome.

Reoviruses have been isolated from the respiratory and intestinal tracts of both ill and well persons worldwide. However, the reoviruses have not been associated with any known disease, hence the name "reo", an acronym for respiratory enteric orphan.

Rotaviruses differ from reoviruses morphologically in that these viruses have two capsids surrounding an icosahedral core. The core does not have spikes at the vertices of the icosahedron. Complete double-shelled rotavirus is about 67 nm in diameter and can be distinguished from reoviruses and orbiviruses by a sharply defined and distinctly characteristic outer rim (Figure 55A). Capsomeres of the outer capsid are encased in a glycoprotein shell. This gives the virus the appearance of a wheel (rota = wheel). Particles without the outer capsid shell are termed single-shelled particles (Figure 55B). These are about 58 nm in diameter. The inner capsid consists of large morphological units arrayed with complex primary and secondary symmetry. As with the outer shell of reoviruses and the inner shell of orbiviruses, these units are formed by sharing of trimeric subunits. The inner capsid is the group specific antigen of the genus Rotavirus. It surrounds an icosahedral core 38 nm in diameter. The inner capsid and core of orbiviruses and rotaviruses have the same morphology. These differ from reovirus cores in that the cores of orbiviruses and rotaviruses do not have spikes at the 12 vertices of the icosahedron as do reoviruses. The main morphologic difference between rotavirus and orbivirus seen by negative stain is that orbiviruses do not have a distinct outer capsomere layer. The surface of orbiviruses frequently appears "fuzzy" (Figure 56).

IEM has been used to detect low levels of rotavirus in stool preparations but, in general, it has not been widely used in the diagnosis of viruses of the Reoviridae. These viruses usually are found in large numbers and are easily recognizable by direct EM.

The replication of reoviruses has some unusual features. Virus is adsorbed and associates with lysosomes. Lysozymal enzymes partly degrade the outer capsid to form a subviral particle containing dsRNA genome segments. The RNAs are transcribed within the subviral particle, each segment coding for a different protein. The RNAs are not released into the cyptoplasm during replication. During morphogenesis reovirus is closely associated with spindle tubules of the mitotic apparatus. Viruses of the Reoviridae generally replicate in the perinuclear area of the cytoplasm in association with granular inclusions or "viral factories" (Figure 57).

Orbivirus and rotavirus are not so closely associated with the spindle tubules as is reovirus. Rotavirus is found in the cytoplasm, often in vacuoles (Figure 58). Maturing orbivirus is often seen with filaments or tubules about the same diameter as the virion (Figure 59). Particles are usually released from the cell by disruption of the cell membrane or by budding. Budding particles may acquire a membranous pseudoenvelope and appear slightly larger than particles released by cell lysis.

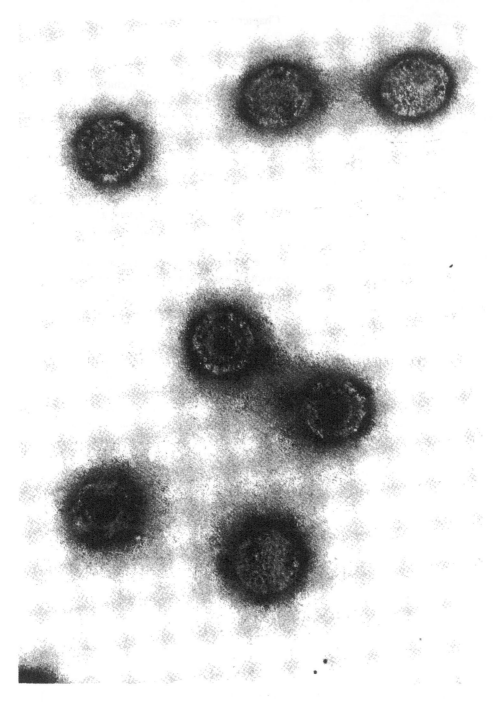

FIGURE 54.   High magnification of reovirus type 3 from culture fluid of infected Hep-2 cells. The inner capsid is clearly visible within particles which have been penetrated by stain. UA stain × 213,300.

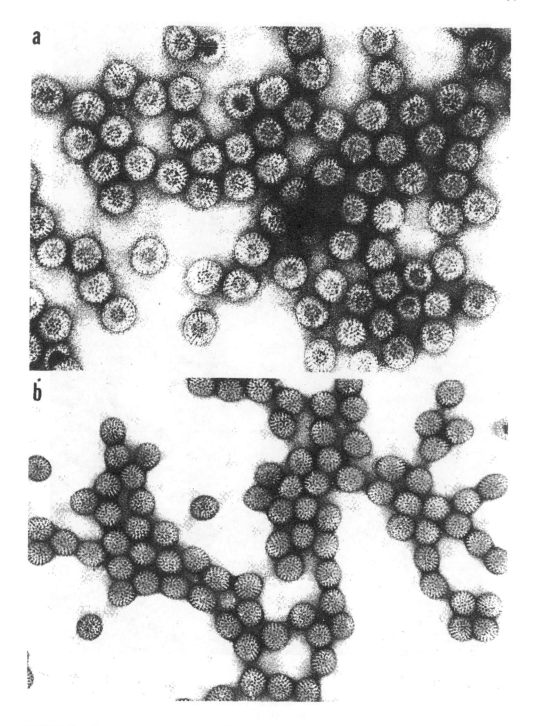

FIGURE 55. Rotavirus propagated in MA104 cells and purified by density gradient centrifugation. (A) The wheel-like structure of double-shelled particles is evident. (B) Ring-shaped capsomeres are seen on the surface of single-shelled particles. UA stain × 127,800.

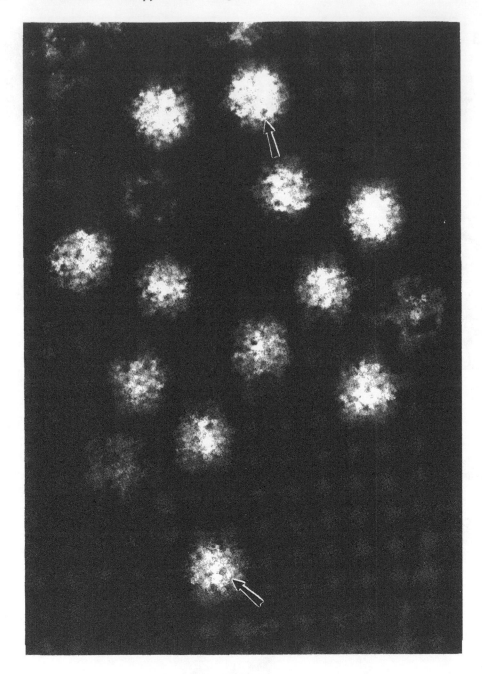

FIGURE 56.   Blue tongue virus negatively stained with PTA. The outer layer of intact particles is "hazy". Ring-shaped capsomeres (arrows) can be seen on the surface of the inner capsid. PTA stain × 248,500. (Courtesy of Frederick Murphy, CDC.)

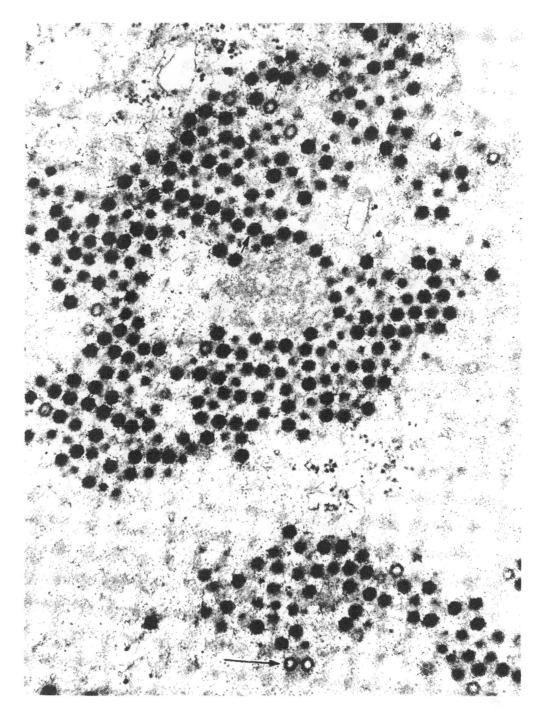

FIGURE 57.   Reovirus in the cytoplasm of infected mouse L cells. Particles have a well defined capsid (short arrow) and an electron dense core. Some empty particles are also present (long arrow). × 73,500. (Courtesy of Alyne Harrison, CDC.)

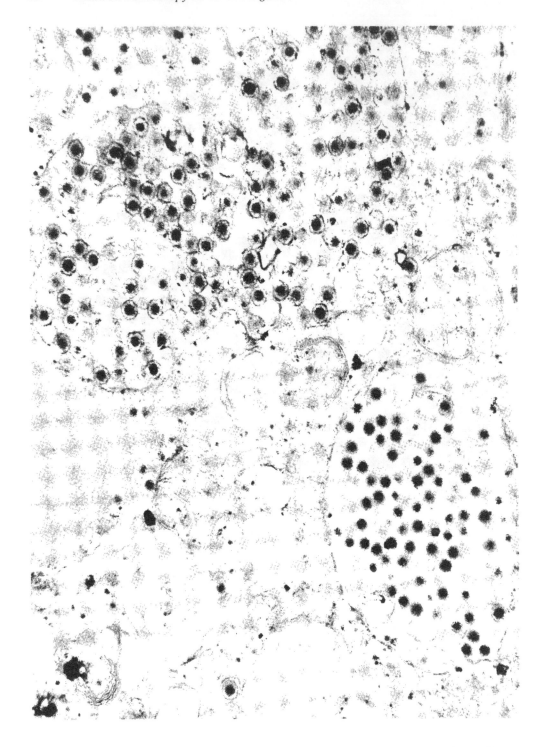

FIGURE 58.   Rotavirus SA-11 in the cytoplasm of infected MA104 cells. Complete and incomplete particles are seen within vacuoles. × 73,500. (Courtesy of Alyne Harrison, CDC.)

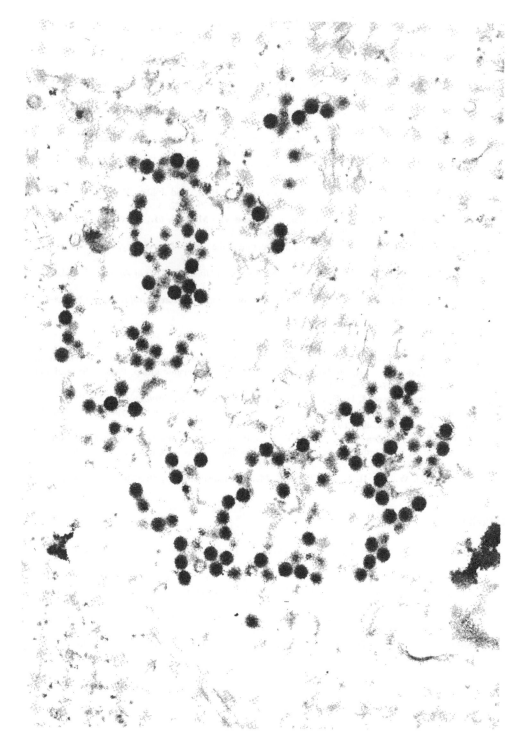

FIGURE 59.   Blue tongue virus particles in the cytoplasm of an infected monkey kidney cell. Particles have an electron dense core and a well defined capsid. A few empty particles and tubules are also present. × 64,600. (Courtesy of Frederick Murphy, CDC.)

REFERENCES

Reovirus

1. **Amano, Y., Katagiri, S., Ishida, N., and Watanabe, Y.,** Spontaneous degradation of reovirus capsid into subunits, *J. Virol.,* 8, 805, 1971.
2. **Gomatos, P. J., Tamm, I., Dales, S., and Franklin, R. M.,** Reovirus type 3: physical characteristics and interaction with cells, *Virology,* 17, 441, 1962.
3. **Joklik, W. K.,** Reproduction of Reoviridae, in *Comprehensive Virology,* Vol. 2, Fraenkel-Conrat, H. and Wagner, R. R., Eds., Plenum Press, New York, 231, 1974.
4. **Jordan, L. E. and Mayor, H. D.,** The fine structure of reovirus, a new member of the icosahedral series, *Virology,* 17, 597, 1962.
5. **Luftig, R. B., Kilham, S., Hay, A. J., and Zweerink, H. J., and Joklik, W. K.,** An ultrastructural study of virions and cores of reovirus type 3, *Virology,* 48, 170, 1972.
6. **Mayor, H. D., Jamison, R. M., Jordan, L. E., and Van Mitchell, M.,** Reoviruses. II. Structure and composition of the virion, *J. Bacteriol.,* 89, 1548, 1965.
7. **Palmer, E. L. and Martin, M. L.,** The fine structure of the capsid of reovirus type 3, *Virology,* 76, 109, 1977.
8. **Shatkin, A. J.,** Viruses containing double-stranded RNA, in *Molecular Basis of Virology,* Fraenkel-Conrat, H., Ed., Reinhold, New York, 1968, 351.
9. **Vasquez, C. and Tournier, P.,** New interpretation of reovirus structure, *Virology,* 24, 128, 1964.

Orbivirus

1. **Borden, E. C., Shope, R. E., and Murphy, F. A.,** Physicochemical and morphologial relationships of some arthropod-borne viruses to bluetongue virus — a new taxonomic group. Physicochemical and serological studies, *J. Gen. Virol.,* 13, 261, 1971.
2. **Els, H. J. and Verwoerd, D. W.,** Morphology of bluetongue virus, *Virology,* 38, 213, 1969.
3. **Martin, S. A. and Zweerink, H. J.,** Isolation and characterization of two types of bluetongue virus particles, *Virology,* 50, 495, 1972.
4. **Murphy, F. A., Borden, E. C., Shope, R. E., and Harrison, A.,** Physiocochemical and morpholoical relationships of some arthropod-borne viruses to bluetongue virus — a new taxonomic group. Electron microscopic studies, *J. Gen. Virol.,* 13, 272, 1971.
5. **Verwoerd, D. W., Els, H. J., DeVilliers, E. M., and Huismans, H.,** Structure of the bluetongue virus capsid, *J. Virol.,* 10, 783, 1972.

Rotavirus

1. **Bridger, J. C. and Woode, G. N.,** Characterization of two particle types of calf rotavirus, *J. Gen. Virol.,* 31, 245, 1976.
2. **Esparza, J. and Gil, F.,** A study on the ultrastructure of human rotavirus, *Virology,* 91, 141, 1978.
3. **Martin, M. L., Palmer, E. L., and Middleton, P. J.,** Ultrastructure of infantile gastroenteritis virus, *Virology,* 68, 146, 1975.
4. **Palmer, E. L., Martin, M. L., and Murphy, F. A.,** Morphology and stability of infantile gastroenteritis virus: comparison with reovirus and bluetongue virus, *J. Gen. Virol.,* 35, 403, 1977.
5. **Roseto, A., Escaig, J., Delain, E., Cohen, J., and Scherrer, R.,** Structure of rotavirus as studied by the freeze-drying technique, *Virology,* 98, 471, 1979.

Chapter 10

TOGAVIRIDAE AND FLAVIVIRIDAE

The togaviruses are a heterogeneous group of viruses which have some common morphological and physicochemical properties. Particles are about 60 to 70 nm in diameter with an envelope surrounding what is thought to be an icosahedral core. Particles have surface projections, are negative stranded, and have a ssRNA genome. The genera of Togaviridae which are medically important include: *Alphavirus* and *Rubivirus*. Alphaviruses are composed of the former group A arboviruses (Sindbis, Venezuelan, Western equine encephalomyelitis, and others). All species of *Alphavirus* multiply in arthropods as well as in vertebrates. *Rubivirus* (rubella virus) does not. Rubella is physicochemically and morphologically similar to other alphaviruses but the virus is not serologically related to other members of the Togaviridae or Flaviviridae. Man is the only known vertebrate host for rubella virus. Rubella virus was one of the first human virus pathogens to be identified by IEM. IEM has also been used to identify unknown alphavirus isolates. The capsid of the alphavirus Sindbis appears to be composed of 32 capsomeres about 14 nm in diameter. The surface subunits are also apparently arrayed in an icosahedral surface lattice, possibly in a T = 4 configuration. Sindbis is the only virus of the Togaviridae which has been studied in detail by negative stain EM. Visualization of togaviruses by negative stain EM has not been satisfactory because particles tend to appear nondescript (Figure 60).

The Flaviviridae comprises the former group B arboviruses (yellow fever, dengue, St. Louis encephalitis, and others) which were formerly members of the Togaviridae. Particles are 40 to 50 nm in diameter (Figure 61) which is somewhat smaller than the range for togaviruses. In addition to size differences, flaviviruses bud from intracytoplasmic membranes, whereas alphavirus buds from the cell surface membrane (Figures 62, 63, and 64.) However, budding flavivirus is not seen often by EM.

Togaviruses and Flaviviruses are positive strand RNA viruses which multiply in the cytoplasm of infected cells. Viral nucleocapsids are associated near cytoplasmic vacuoles. These acquire envelopes while passing into vacuoles, or as particles are extruded from the plasma membrane. Elongated forms of Togaviruses may also be seen budding from the surface of infected cells. Flaviviruses are usually seen in cytoplasmic vacuoles and are probably released through canaliculi connecting the endoplasmic reticulum with the plasma membrane or perhaps the cellular secretory mechanism.

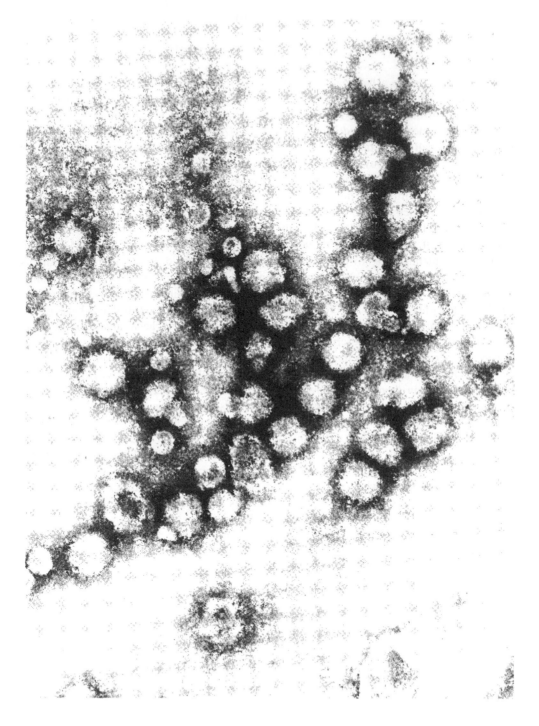

FIGURE 60.   Rubella virus negatively stained with PTA. Surface structure is not distinctive but close examination shows fuzzy surface projections. × 122,000.

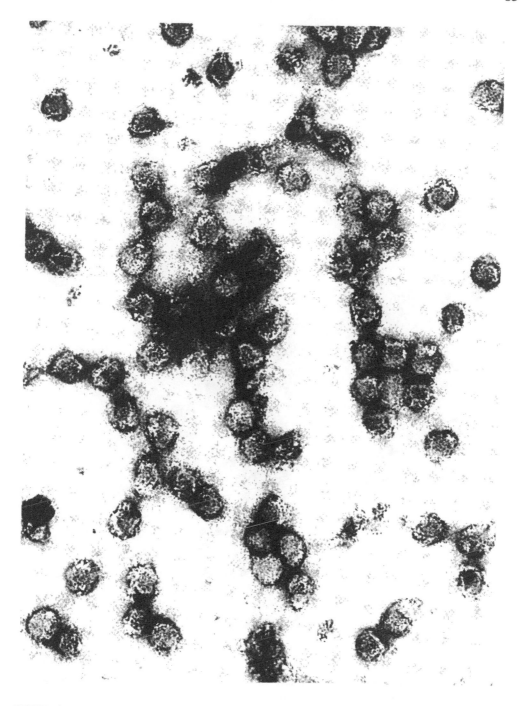

FIGURE 61. Vaccine strain of the flavivirus, yellow fever virus, purified by density gradient centrifugation and negatively stained with UA. Particles are morphologically indistinct but are compact and about the same shape and size. × 195,325.

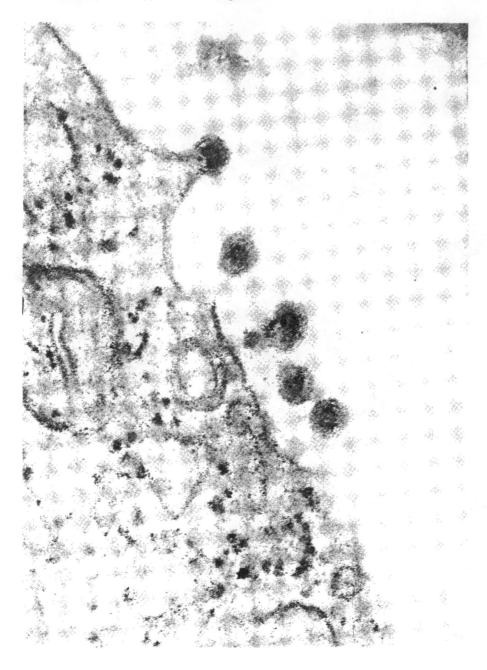

FIGURE 62.   Rubella virions budding from marginal membrane of an infected BHK-21 cell. Membrane appears contiguous with outer layer of virus particles. × 154,460. (Courtesy of Frederick Murphy, CDC.)

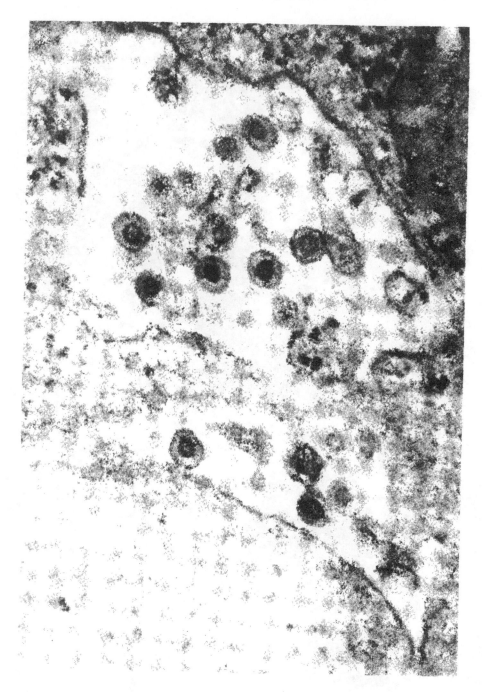

FIGURE 63.   Rubella virus budding into an extracellular space of an infected BHK21 cell. × 154,460.
(Courtesy of Frederick Murphy, CDC.)

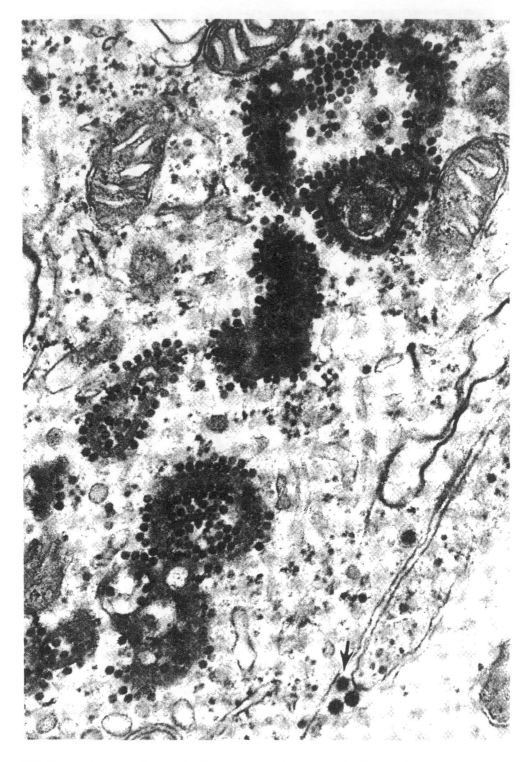

FIGURE 64.   Eastern equine encephalitis virus in mouse brain. Nucleoids 20nm in diameter are seen within a cytoplasmic matrix and upon complex membranous structures. Enveloped extracellular particles with electron dense centers and ragged surfaces are also evident (arrow, lower right). × 100,000. (Courtesy of Frederick Murphy, CDC.)

REFERENCES

1. **Acheson, N. H. and Tamm, I.**, Purification and properties of Semliki Forest virus nucleocapsids, *Virology*, 41, 306, 1970.
2. **Berge, T. O.**, International catalogue of arboviruses including certain other viruses of vertebrates, U.S. Dept. of Health, Education, and Welfare, DHEW Publication No. (CDC) 75-8301, Washington, D.C., 1975.
3. **Brown, D. T. and Gliedman, J. B.**, Morphological variants of Sindbis virus obtained from infected mosquito tissue culture cells, *J. Virol.*, 12, 1534, 1973.
4. **Brown, D. T., Waite, M. R. F., and Pfefferkorn, E. R.**, Morphology and morphogenesis of Sinbis virus as seen with freeze-etching techniques, *J. Virol.*, 10, 524, 1972.
5. **Enzmann, P. J. and Weiland, F.**, Studies on the morphology of alphaviruses, *Virology*, 95, 501, 1979.
6. **Horzinek, M. C.**, The structure of togaviruses, *Prog. Med. Virol.*, 16, 109, 1973.
7. **Horzinek, M. and Mussgay, M.**, Studies on the nucleocapsid structure of a group A arbovirus, *J. Virol.*, 4, 514, 1969.
8. **Murphy, F. A., Harrison, A. K., Gary, G. W., Jr., Whitfield, S. G., and Forrester, F. T.**, St. Louis encephalitis virus infection of mice. Electron microscopic studies of central nervous system, *Lab. Invest.*, 19, 652, 1968.
9. **Norrby, E.**, Rubella virus, *Virol. Mongr.*, 7, 115, 1969.
10. **Pfefferkorn, E. R. and Shapiro, D.**, Reproduction of togaviruses, in *Comprehensive Virology*, Vol. 2, Fraenkel-Conrat, H. and Wagner, R. R., Eds., Plenum Press, New York, 1974, 171.
11. **Von Bonsdorff, C. H. and Harrison, S. C.**, Sindbis virus glycoproteins form a regular icosahedral surface lattice, *J. Virol.*, 16, 141, 1975.

Chapter 11

RETROVIRIDAE

The Retroviridae family of viruses consists of a large number of RNA tumor viruses and related viruses which have a virion-associated reverse transcriptase. Retro, the vernacular for backwards, refers to this endogenous transcriptase. Many of these retroviruses are known to cause leukemias, lymphomas, and sarcomas, in various animal species. The retroviruses that cause tumors are classified as a subfamily, Oncovirinae, and are separated into genera as retrovirus type B, C and D on the basis of (1) morphology in thin section EM, (2) antigenic relatedness, (3) physicochemical properties, (4) nucleic acid hybridization, and (5) genome organization. The genera are further divided into subgenera on the basis of the host and group-specific antigenic determinant of the retrovirus internal group specific "gag" protein. There are two other subfamilies, the subfamily Lentivirinae (Maedi/Visna group), and the subfamily Spumavirinae (foamy virus group).

Retroviruses are roughly spherical, measuring about 85 to 130 nm in diameter and have a lipid envelope with spike or knob-like surface projections. The envelope encloses a core which, in turn, encloses helical RNP. The genome is single-stranded RNA. Virions are sensitive to osmotic pressure brought about by specimen preparation and are fragile after purification.

In the early 1980s retroviruses were isolated from patients with T cell malignancies and from patients with the acquired immunodeficiency syndrome (AIDS) or persons at risk for AIDS. These viruses have been designated as human T cell leukemia virus (HTLV) types I, II and human immunodeficiency virus (HIV). HTLV I and II have the same morphology but are different from HIV (Figure 65). HTLV-I is a type C retrovirus presumed to be the cause of adult T cell leukemia. HTLV-II is also a type C retrovirus which was isolated from lymphocytes of a patient with a T cell variant of hairy cell leukemia and from a hemophiliac with pancytopenia. Both HTLV-I and II are lymphotropic and can transform cells in vitro.

HIV has been isolated from a large number of patients with AIDS. It is a T cell cytopathic rather than transforming virus. HIV isolates are identical in morphology to the retrovirus which was first isolated from a homosexual with an AIDS related syndrome by workers at the Pasteur Institute in France and termed lymphadenopathy associated virus (LAV) and to human T lymphotropic virus III (HTLV-III) subsequently isolated from AIDS patients in the U.S. The designation HIV was suggested by the ICTV as a vernacular name to refer to isolates which are the same as the original LAV and HTLV-III isolates. HIV closely resembles viruses of the subfamily Lentivirinae, having a bar or conical shaped, electron dense eccentric nucleoid like members of this subfamily (Figure 66). The virus is markedly similar in morphology to equine infectious anemia and Visna viruses and is different in morphology from HTLV-I and HTLV-II (Figures 67, 68, and 69).

Simian retroviruses are endemic in several species of monkeys. Viruses related to HTLV-I and possibly HTLV-II infect a variety of old world monkeys and are associated with lymphoma. These are type C oncoviruses which have the same morphology as HTLV-I. Some simian retroviruses are also similar to AIDS retroviruses isolated in West Africa. Simian lymphotropic virus type III (STLV-III MAC) causes AIDS in captive macaque monkeys. Another strain of the virus (STLV-III GMK) is found in captured wild African green monkeys. These viruses have different envelope proteins than HIV so the serological cross reactivity between STLV-III and HIV is not strong.

STLV-III(MAC) is strongly cross reactive and has similar envelope proteins as another virus isolated from humans with AIDS in West Africa. This human virus has been termed lymphadenopathy-AIDS virus type II (LAV-II). LAV-II is more reactive with antiserum to STLV-III MAC than to antiserum to prototype HIV. LAV-II is cytopathic to T4 lymphocytes

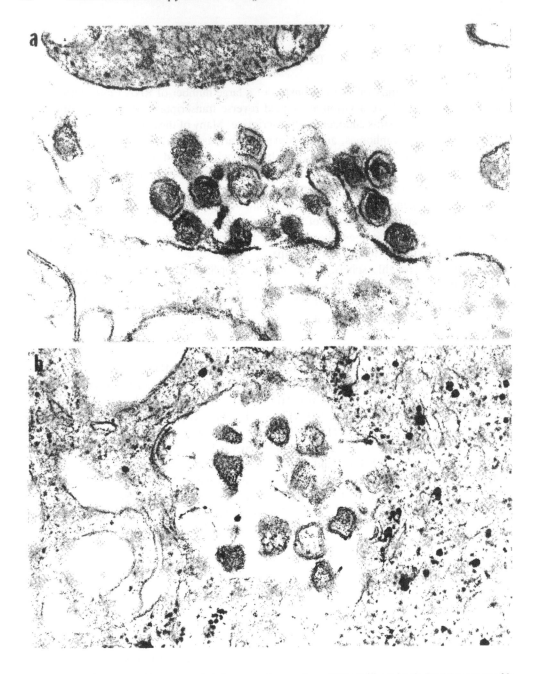

FIGURE 65.   (A) HTLV-I in an extracellular space of a transformed T4 cell line (MT-1). Particles are roughly spherical and have a core which almost completely fills the particle. × 69,750. (B) HTLV-II in a cytoplasmic vacuole of a transformed T4 cell line. Particles are similar in morphology to HTLV-I. These viruses are morphologically distinct from AIDS virus (HIV). × 69,750.

in vitro. Another retrovirus termed human lymphotropic virus type IV (HTLV-IV) was isolated from persons in West Africa without AIDS. It is not cytopathic in vitro but strongly cross reacts with STLV-III GMK. These human and simian retroviruses have a similar morphology. The viruses have a bar or conical shaped nucleoid, are roughly round, and have very distinctive surface projections which are always visible in thin sectioned preparations (Figure 70). HIV also has surface projections but these are not always clearly visible. The spikes are seen only rarely in negatively stained preparations (Figure 71).

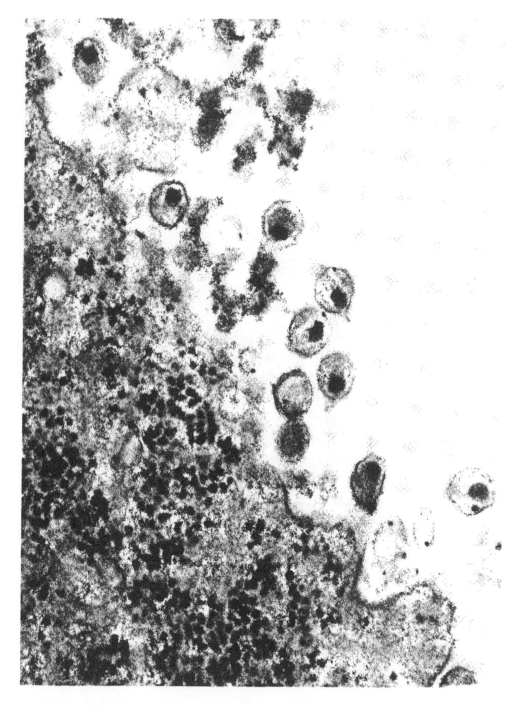

FIGURE 66.   Human immunodeficiency virus (HIV) in an extracellular space of an infected T4 lymphocyte. Particles are roughly spherical and have a distinctive electron dense nucleoid surrounded by a less well-defined membrane. Nucleoids are often seen as bar- or cone-shaped. × 92,680.

The human tumor viruses and HIV replicate by a similar mechanism. To initiate infection these retroviruses adsorb to specific cell receptors and penetrate the plasma membrane. Shortly after infection a DNA intermediate in the replication of viral RNA is synthesized. A minus strand complementary to viral RNA is made by reverse transcriptase. A plus DNA strand complementary to the minus DNA strand is also made. The DNA moves to the nucleus

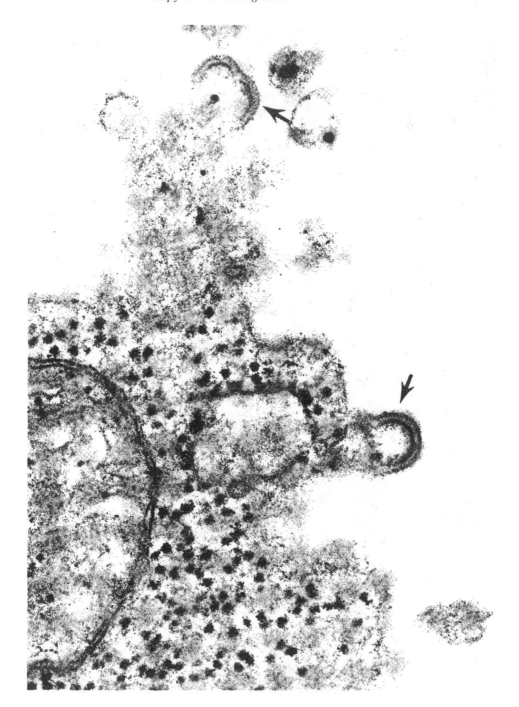

FIGURE 67.   Human immunodeficiency virus (HIV) budding from a T4 lymphocyte. The nascent particles have electron dense crescent-shaped nucleoids and fuzzy surface projections. × 190,000.

and integrates with cellular DNA as provirus. Progeny viral RNA is generated by transcription of the integrated provirus. Envelope and core proteins along with viral RNA are assembled directly beneath the plasma membrane. The nucleocapsid assembles as nascent virus buds from the plasma membrane. No intermediate cytoplasmic forms are involved in the replication of type C retroviruses or with HIV. However, other retroviruses may have an intermediate

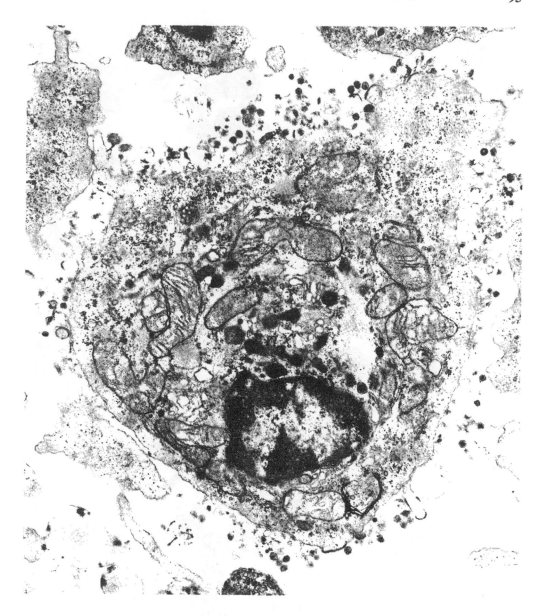

FIGURE 68. Low magnification of a thin section of a human T4 lymphcyte infected with both HTLV-1 and HIV. Virus particles appear as roughly round structures around the periphery of the cell. × 20,260.

cytoplasmic form. A schematic showing the overall replicative events in HIV morphogenesis is shown in Figure 72 and a schematic of the comparative morphogenesis of retrovirus types B, C, D, Lentivirus, and HIV is presented in Figure 73. Figure 73 is a diagramatic representation of the morphological features of some mammalian retroviruses as indicated in the following description:

1. Top. Type B virus — mouse mammary tumor virus (MMTV). Particles are thought to derive from intracytoplasmic type A particles (Figure 74) which are located inside the cisternae of the endoplasmic reticulum. Mature type B particles have prominent surface projections and electron dense eccentric nucleoid(s). Budding particles have an electron lucent core.

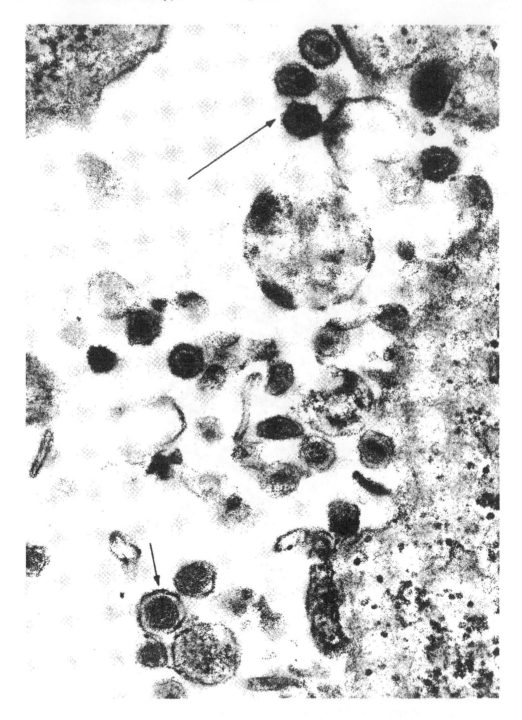

FIGURE 69.   Thin section of a lymphocyte culture established from cells of an AIDS patient from Central Africa. This cell line replicates both HTLV-I (short arrow) and HIV (long arrow). × 109,382. (Courtesy of Cynthia Goldsmith, CDC.)

2.   Left. Type C virus — feline leukemia virus (FLV), human T cell leukemia virus types I, II. Particles do not have an intracytoplasmic nucleocapsid precursor. Particles bud from the plasma membrane or into cytoplasmic vacuoles. The prominent feature of nascent C particles is an electron dense crescent-shaped nucleoid. The surface of budding particles is often fuzzy and short knobs may be delineated. Immature C

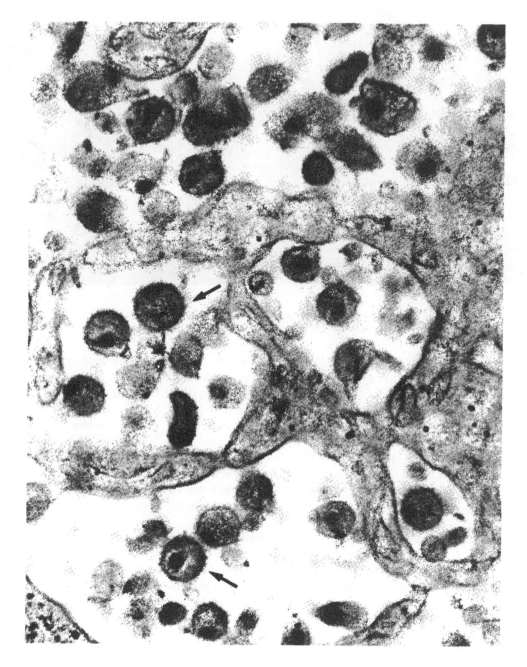

FIGURE 70. Lymphadenopathy-AIDS virus type II in an intracellular space of a lymphocyte. LAV-II has prominent surface projections (arrows) which distinguish it from prototype LAV-I (now termed HIV). × 103,700.

particles have an electron-lucent center. The nucleoid of mature type C particles is centric and loosely arrayed. It fills most of the mature particle but does not appear to come into contact with the inner surface of the viral envelope.

3.   Right. Type D virus — Mason-Phizer monkey virus (MPMV). Particles are thought to derive from intracytoplasmic A particles. Budding particles resemble type C particles but do not have a crescent-shaped nucleoid. Mature particles have a centric electron dense nucleoid which is sometimes bar-shaped.

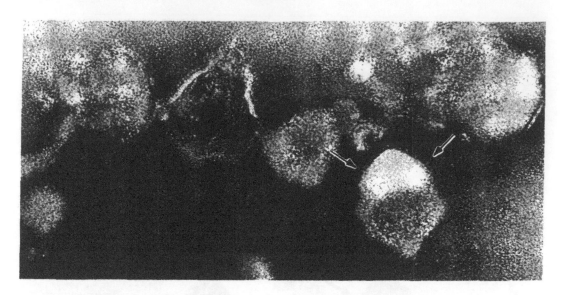

FIGURE 71.   Negative stain of HIV (prototye LAV-I) from the culture fluid of infected lymphocytes. Virus has short surface projections (arrows) and an envelope. UA stain × 195,325. (Courtesy of Cornelia Greene, CDC.)

4.   Lower right. Lentivirus — (Maedi/Visna virus). Particles form by budding from the plasma membrane and nascent forms may resemble type C particles. Intermediate ring forms are common. Mature particles have electron dense eccentric nucleoids.

5.   Lower left. HIV (LAV-HTLV III prototypes) and other AIDS isolates. Particles do not appear to have an intracytoplasmic precursor. Virus forms by budding from the plasma membrane and appears similar in morphology to lentiviruses. Nucleoids are eccentric and electron dense. The nucleoids have bar or conical shapes which are often seen as round in thin section. ''Tear drop''-shaped particles with bar-shaped nucleoids are frequently seen in lymphocytes infected with this virus. The nucleoid is enclosed by a membrane and does not come into contact with the inner viral envelope.

This diagram points out some of the main features of retroviruses as seen in thin section. Type B, C, D, and spumaviruses (Figure 75) can be identified by morphological features of mature particles. However, lentiviruses and HIV have the same general morphology. Viruses with the same morphology as HIV have also been isolated from the lymphocytes of seemingly healthy monkeys. It is probable that other human retroviruses will be unveiled as work on the viruses causing T cell leukemia and AIDS intensifies.

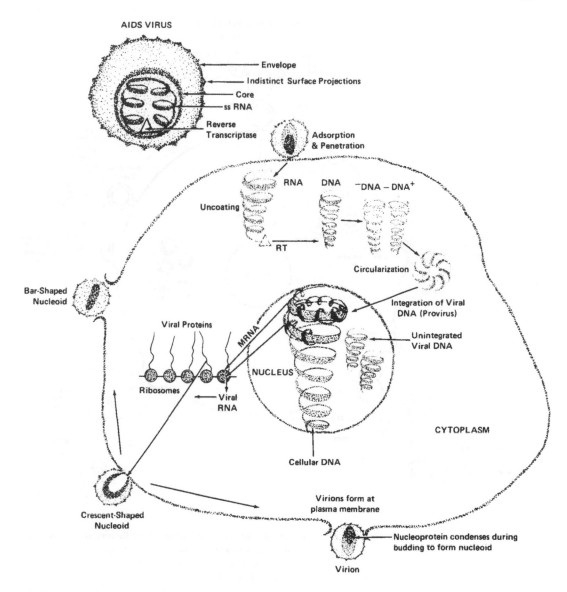

FIGURE 72. Schematic showing the overall events occurring during the replication of HIV. DNA complementary to viral RNA is made by reverse transcriptase. It then integrates as provirus to direct the synthesis of virus progeny. Particles are formed at the plasma membrane. There is no evidence for a nucleocapsid precursor.

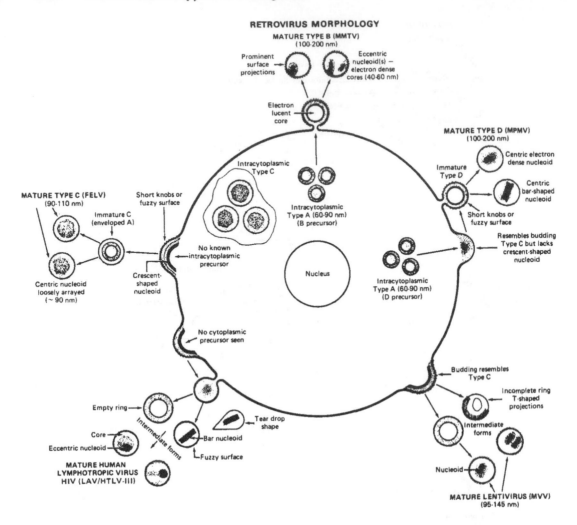

FIGURE 73.   Schematic summary showing an overview of the replication of some viruses of the Retroviridae. Types B, C, and D viruses have a distinctive morphology. Viruses can usually be grouped by the arrangement and location of the formation of nucleoid(s). Viruses of the Lentivirinae subfamily and HIV virus have the same overall morphology. These viruses can be recognized by the presence of an electron dense bar- or cone-shaped nucleoid.

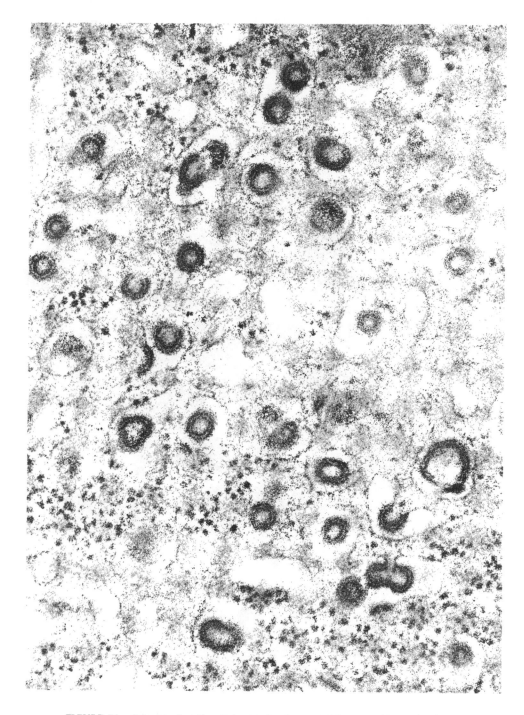

FIGURE 74. Intracytoplasmic type A retrovirus in a neuroblastoma cell. × 102,600.

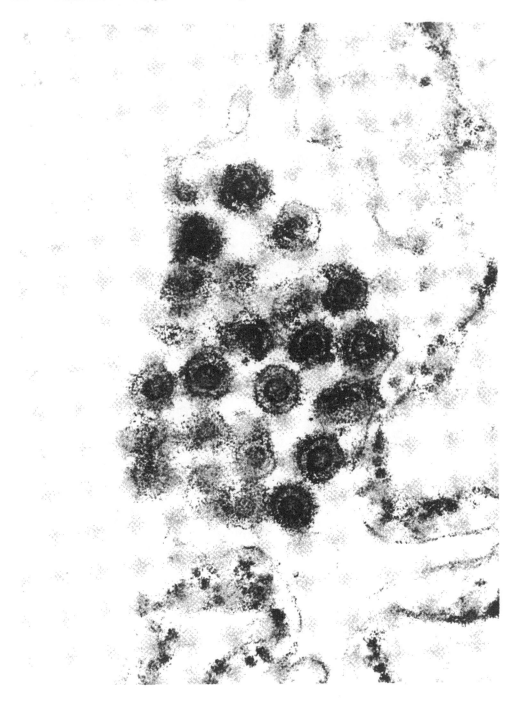

FIGURE 75.   Foamy virus in an extracellular space of a monkey kidney cell. Particles have a "doughnut" shape which is characteristic for viruses of the Spumavirinae subfamily. × 155,000. (Courtesy of Alyne Harrison, CDC.)

REFERENCES

1. **Barre-Sinoussi, F., Chermann, J. C., Rey, F., Nugeyure, M. T., Chamaret, S., Gruest, J., Dauguet, C., Axler-Blin, C., Vezinet-Brun, F., Rouzioux, C., Rozenbaum, M., and Montagnier, L.,** Isolation of T-lymphotropic retrovirus from a patient at risk for acquired immune deficiency syndrome (AIDS), *Science,* 220, 868, 1983.

2. **Bernhard, W.,** The detection and study of tumor viruses with the electron microscope, *Cancer Res.,* 20, 712, 1960.

3. **Bonar, R. A., Heine, U., Beard, D., and Beard, J. W.,** Structure of BAI strain A (myeloblastosis) avian tumor virus, *J. Natl. Cancer Inst.,* 30, 949, 1963.

4. **Clavel, F., Guetard, A., Brun-Vizinet, F., Chamaret, S., Rey, M. A., Santos-Ferreira, M. O., Laurent, A. G., Dauguet, C., Katlama, C., Rouzioux, C., Klatzmann, D., Champalimaud, J. L., and Montagnier, L.,** Isolation of a new human retrovirus from West African patients with AIDS, *Science,* 233, 343, 1986.

5. **Chopra, H. C. and Mason, M. M.,** A new virus in a spontaneous mammary tumor of a rhesus monkey, *Cancer Res.,* 30, 2081, 1970.

6. **Dalton, A. J., Potter, M., and Mervin, R. M.,** Some ultrastructural characteristics of a series of primary and transplanted plasma-cell tumors of the mouse, *J. Natl. Cancer Inst.,* 26, 1221, 1961.

7. **Gallo, R. C., Salahuddin, S. Z., Popovic, M., Shearer, G. M., Kaplan, M., Haynes, B. F., Polkev, T. J., Redfield, R., Olesbe, J., Safai, B., White, G., Foster, P., and Markam, P. O.,** Frequent detection and isolation of cytopathic retroviruses (HTLV-III) from patients with AIDS and at risk for AIDS, *Science,* 224, 500, 1984.

8. **Homma, T., Kanki, P. J., King, N. W., Hunt, R. D., O'Connell, J. J., Letvin, N. L., Daniel, M. D., Desrosiers, R. C., Yang, C. S., and Essex, M.,** Lymphoma in macaques: association with exposure to virus of human T lymphotropic family, *Science,* 225, 716, 1984.

9. **Kanki, P. J., Kurth, R., Becker, W., Dreesman, G., and Essex, M.,** Antibodies to simian T-lymphotropic retrovirus type III in Africa green monkeys and recognition of STLV III viral proteins by AIDS and related sera, *Lancet,* 1, 1330, 1985.

10. **Kalyanaraman, V. S., Sarngadharan, M. G., Miyoshi, I., Robert-Guroff, M., Miyoshi, I., Blayney, D., Golde, D., and Gallo, R. C.,** A new subtype of human T-cell leukemia virus (HTLV-II) associated with a T-cell variant of hairy-cell leukemia, *Science,* 218, 571, 1983.

11. **Levy, J. A., Hoffman, A. D., Kramer, S. M., Landis, J. A., Shimabykuro, J. M., and Oshiro, L. S.,** Isolation of lymphotropic retroviruses from San Francisco patients with AIDS, *Science,* 225, 840, 1984.

12. **Nermut, M. V., Herman, F., and Schaffer, W.,** Properties of mouse leukemia viruses. III. Electron microscopical appearance as revealed after conventional preparation techniques as well as freeze drying and freeze etching, *Virology,* 49, 345, 1972.

13. **Poiesz, B. J., Ruscetti, F. W., Gazdar, A. F., Bunn, P. A., Minna, J. A., and Gallo, R. C.,** Detection of type C retrovirus particles from fresh and cultured lymphocytes of a patient with cutaneous T-cell lymphoma, *Proc. Natl. Acad. Sci. USA,* 77, 7415, 1980.

14. **Tooze, J.,** RNA tumour viruses: morphology, composition and classification, in *The Molecular Biology of Tumor Viruses,* Cold Spring Harbor Laboratory, New York, 1973, 502.

Chapter 12

ORTHOMYXOVIRIDAE

The most important members of the Orthomyxoviridae family are influenza virus types A and B which have been given the generic status of *Influenza*. *Influenza* type A has been isolated from humans, swine, horses, and birds, whereas type B has been isolated only from humans. *Influenza* viruses are typed as A or B on the basis of the antigenic specificity of their internal ribonucleoprotein and matrix protein. External proteins (the hemagglutinin [HA] and the neuraminidase [NA]) comprise additional antigens of *Influenza* type A and B viruses. A number of subgroups have been recognized among hemagglutinin and neuraminidase antigens of *Influenza* A virus, and to a lesser extent, antigenic variation also occurs in the external antigens of *Influenza* B virus. Strains of influenza A or B viruses are identified on the basis of differences in the antigenic subgroups of hemagglutinin and neuraminidase. These antigens are capable of independent variation from each other.

Viruses of freshly isolated strains of *Influenza* are often heterogeneous in size and shape. Filamentous forms sectioned in various planes are usually seen in thin sectioned preparations (Figure 76). Influenza can, nevertheless, be differentiated from other viruses by the presence of evenly spaced spikes covering the entire virus surface. These spikes may be seen projecting end-on toward the viewer or as a "fringe" surrounding an electron-lucent center. They project about 9 nm from the surface of intact particles, are 4 to 8 nm wide, and are spaced 7 to 8 nm apart (See Figure 12, Chapter 1) . The spike layer contains the external influenza antigens: hemagglutinin and a virus-coded enzyme which has been identified in *Influenza* A and B as a neuraminidase. The hemagglutinin attaches to target cells, and agglutinates red blood cells from a variety of animal species. Neuraminidase hydrolyzes neuraminic acid residues from mucoproteins and allows virus to elute from cell receptors.

Beneath the spike layer is a characteristic lipid-containing envelope enclosing a protein matrix and a helical nucleocapsid formed of protein and single-stranded RNA. A viral RNA-dependent RNA polymerase is associated with the nucleocapsid of *Influenza* A virus, and probably with the nucleocapsids of types B and C as well. Viruses which require an RNA-dependent RNA polymerase for replication are termed negative stranded viruses.

The envelope of influenza virus is 6 to 10 nm thick. When this layer is disrupted or penetrated by stain, a nucleocapsid can sometimes be seen as folded parallel bands (See Figure 9, Chapter 1). Free-lying nucleocapsids can be obtained by treating virions with detergent. They appear as small supercoiled helices of varying size. There are eight RNP segments, each of which codes for a separate protein. The viral RNA is not infectious.

Influenza C virus (Figure 77), which has not been given the status of a genus of the Orthomyxoviridae family, may appear to be identical in morpology to *Influenza* A or B, or may be stained so that a hexagonal structure of the virus surface is clearly evident. Usually both types of particles are seen in the same field.

Figure 78 is a schematic representation of the steps in the synthesis of influenza. The site of influenza viral RNA transcription and replication is unknown but is thought to involve the nucleus. Influenza hemagglutinin spikes attach to host-cell receptors and particles fuse with the cytoplasmic membrane. The viral envelope is stripped away and the nucleocapsids are free in the cytoplasm. These migrate to the nucleus where viral RNA transcription is thought to occur. The RNA-dependent RNA polymerase within the particle transcribes viral RNA segments into mRNAs. Nucleocapsid protein is synthesized in the cytoplasm and is probably transported to the nucleus to form nucleocapsids. These then migrate to the cell membrane. The matrix protein becomes associated with the inner surface of the plasma membrane, and this area also incorporates HA and NA protein. Nucleocapsids migrate to these areas of the membrane to form virus particles. Other negative stranded viruses replicate

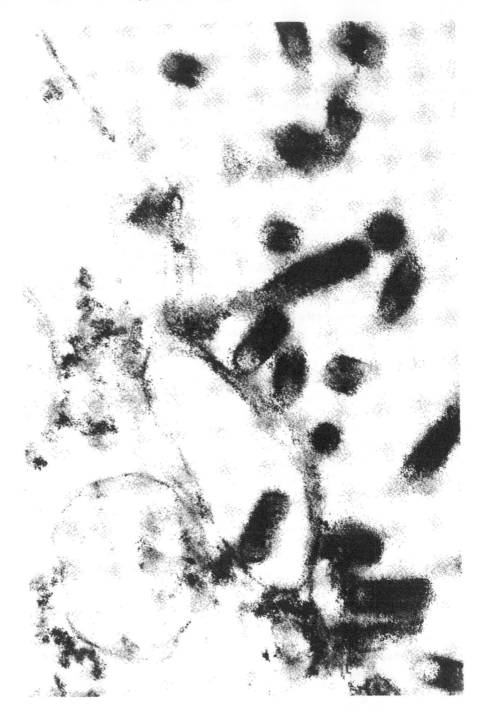

FIGURE 76.   Thin section of a monkey kidney cell infected with influenza virus. This field contains particles sectioned in a plane such that virus is cylindrical. × 155,000.

in a similar sequence of steps involving: (1) adsorption, (2) penetration, (3) synthesis of viral complementary mRNA, (4) translation, and (5) replication of viral RNA and eventually assembly of the translation products into virions as particles bud from cell membranes. The replicative process takes place primarily in the cytoplasm.

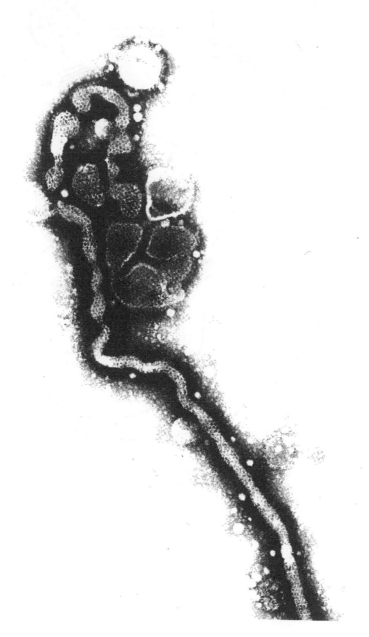

FIGURE 77. Influenza type C negatively stained with UA. The surface of this virus is composed of hexagonal arrays of subunits. × 75,330.

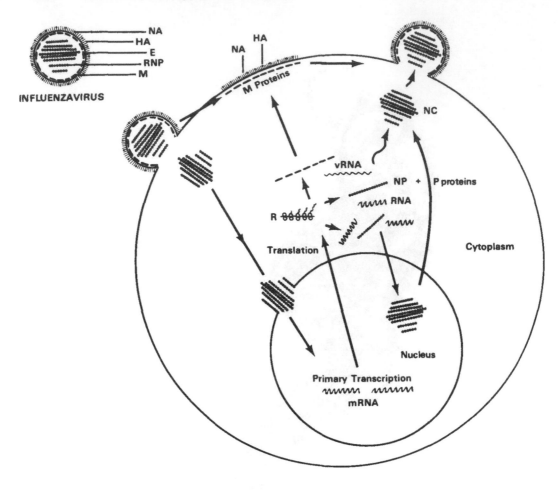

FIGURE 78. Schematic representing the replication of influenza virus. This is an example of the replication of negative stranded enveloped RNA viruses. Virions are formed at the plasma membrane. There is no evidence for a cytoplasmic precursor.

REFERENCES

1. **Choppin, P. W. and Compans, R. W.,** The structure of influenza virus, in *The Influenza Viruses and Influenza,* Kilbourne, E. D., Ed., Academic Press, New York, 1975, 15.
2. **Choppin, P. W., Murphy, J. S., and Stoeckenius, W.,** The surface structure of influenza virus filaments, *Virology,* 13, 548, 1961.
3. **Compans, R. W. and Choppin, P. W.,** Orthomyxoviruses and paramyxoviruses, in *Ultrastructure of Animal Viruses and Bacteriophages: An Atlas,* Dalton, J. A. and Haguenau, F., Eds., Academic Press, New York, 1972, 213.
4. **Duesberg, P. H.,** Distinct subunits of the ribonucleoprotein of influenza virus, *J. Mol. Biol.,* 42, 485, 1969.
5. **Horne, R. W., Waterson, A. P., Wildy, P., and Farnham, A. E.,** The structure and composition of the myxoviruses. I. Electron microscope studies of the structure of the myxovirus particles by negative staining techniques, *Virology,* 11, 79, 1960.
6. **Hoyle, L., Horne, R. W., and Waterson, A. P.,** The structure and composition of the myxoviruses. II. Components released from the influenza virus particle by ether, *Virology,* 13, 448, 1961.
7. **Laver, W. G. and Valentine, R. C.,** Morphology of the isolated hemagglutinin and neuraminidase subunits of influenza virus, *Virology,* 38, 105, 1969.
8. **Martin, M. L., Palmer, E. L., and Kendal, A. P.,** Lack of characteristic hexagonal surface structure on newly isolated influenza C virus, *J. Clin. Microbiol.,* 6, 84, 1977.
9. **Pons, M. W., Schulze, I. T., and Hirst, G. K.,** Isolation and characterization of the ribonucleoprotein of influenza virus, *Virology,* 39, 250, 1969.
10. **Tiffany, J. M. and Blough, H. A.,** Models of structure of the envelope of influenza virus, *Proc. Natl. Acad. Sci. U.S.A.,* 65, 1105, 1970.

REFERENCES

Chapter 13

PARAMYXOVIRIDAE

The Paramyxoviridae family consists of three genera important to humans: *Paramyxovirus*, comprising parainfluenza virus types 1 through 4 (type 4 has subtypes A and B), the Newcastle disease virus group, and mumps virus; *Morbillivirus*, comprising the measles (rubeola) virus group; and *Pneumovirus*, which includes respiratory syncytial virus and related agents. *Paramyxovirus* and *Morbillivirus* are identical in morphology. *Pneumovirus*, although having some morphological features in common with other Paramyxoviridae, differs enough to be distinguishable from them by EM.

The *Paramyxovirus* genus encompasses a number of viruses which cause clinical disease in humans. Significant serological cross-reactivity occurs among viruses within the genus, but there is no detectable cross-reactivity with *Morbillivirus*, *Pneumovirus*, or members of the family Orthomyxoviridae. All *Paramyxovirus* species except respiratory syncytial virus have three type-specific antigens; a hemagglutinin surface antigen, a neuraminidase surface antigen, and an internal nucleocapsid antigen. However, for mumps virus the hemagglutinin is also commonly called the V (viral) antigen and the ribonucleoprotein antigen is referred to as the S (soluble) antigen.

The *Paramyxovirus* and *Morbillivirus* members are quasi-spherical and average about 125 to 250 nm in diameter; however, some individual particles may be very small (100 nm) or very large (approaching 1 μm). The nucleocapsid is surrounded by an envelope about 10 nm thick. The envelope is covered with clearly seen regularly arrayed spikes consisting of the hemagglutinin and neuraminidase antigens (Figure 79). The nucleocapsid, unlike that of the Orthomyxoviruses, is a continuous helix. It has a width of approximately 18 nm and a central hole of about 4 nm; a periodicity of 4 nm has been determined for the serrations. This structure is quite distinctive when seen by negative-contrast electron microscopy, resembling a "herringbone" pattern (Figure 80, also see Figure 7A, Chapter 1), and provides a means of distinguishing *Paramyxovirus* and *Morbillivirus* from Influenza A or B viruses.

The *Morbillivirus* measles, and related viruses, are identical in morphology to members of the *Paramyxovirus* genus when viewed by negative-contrast electron microscopy. However, they do not have antigens in common with the paramyxoviruses and have several distinct features. Measles virus possesses a hemagglutinin corresponding to its surface projections, but the projections do not have neuraminidase activity. Further, measles virus contains a lipid soluble factor which hemolyzes monkey erythrocytes and induces giant-cell formation of cultured cells. Antigenic variations among measles virus strains have not been detected.

Respiratory syncytial virus (RSV), the type species of the *Pneumovirus* genus, causes disease mostly in young children. It affects the entire respiratory tract, producing a variety of syndromes such as the common cold, laryngotracheobronchitis (croup), and a host of other maladies. It was named RSV because it characteristically produces large syncytial masses in tissue culture. RSV has a number of properties that separate it from *Paramyxovirus* and measles virus. It possesses surface projections which are club-like in appearance and are regularly spread, but it lacks hemagglutinin, neuraminidase, or hemolytic activity. RSV readily forms filaments and is often very pleomorphic. Most particles are roughly spherical and average 90 to 130 nm in diameter. The nucleocapsid is about 14 nm wide, with a periodicity of serrations of about 7 nm.

The multiplication cycle is generally the same for each paramyxovirus (Figure 81). The basic biochemical events follow those of other negative stranded RNA viruses which contain an RNA dependent RNA polymerase. Particles form at the surface of infected cells by budding from the plasma membrane (Figures 82, 83, and 84). It is often possible to see

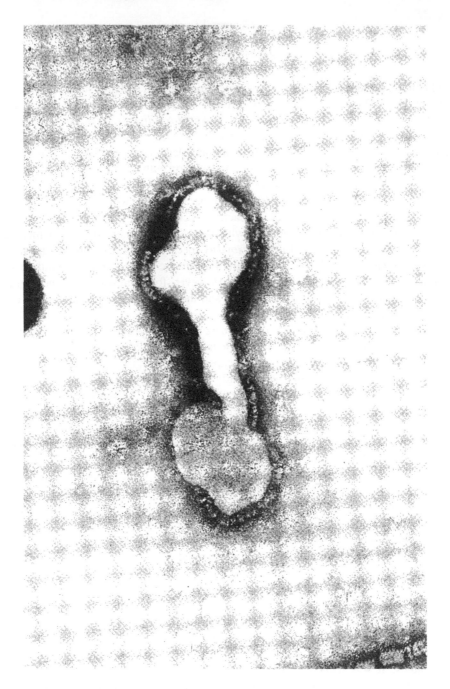

FIGURE 79.   Parainfluenza virus negatively stained with UA. The surface of this virus is covered with fine projections. × 195,325.

nucleocapsids in cross section under a region of membrane (Figures 85 and 86), and filamentous forms are common. Large accumulations of nucleocapsids may occur in the cytoplasm of infected cells.

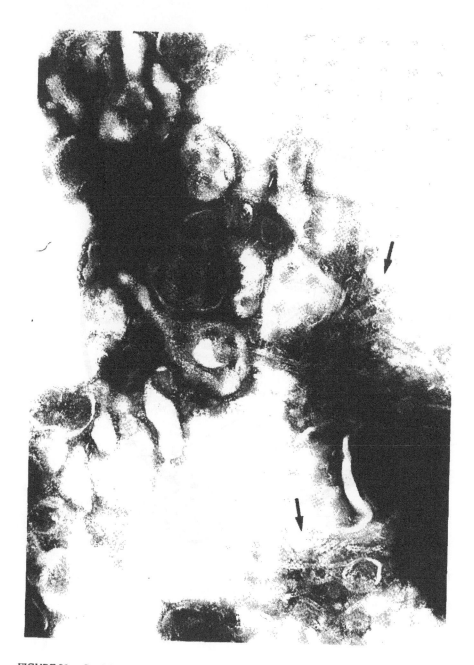

FIGURE 80. Sendai virus, a paramyxovirus, negatively stained with PTA. Free lying nucleocapsids with a "herringbone" structure (arrows) are evident. These structures are also seen within particles which have been penetrated by stain. × 89,460.

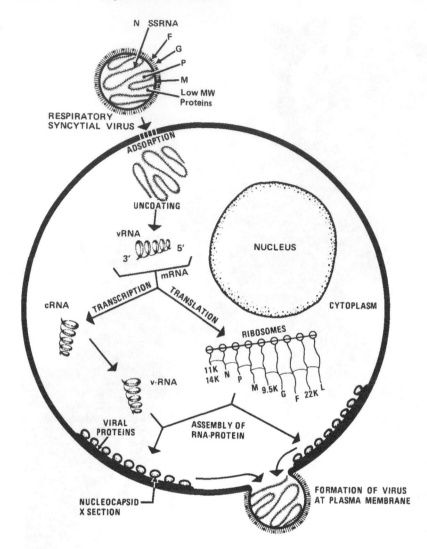

FIGURE 81.   Schematic drawing showing an overall view of the replication of respiratory syncytial virus, a *Paramyxovirus*. Virions are formed at the plasma membrane after nucleocapsids are formed just under the membrane. Particles become complete upon egress from the cell.

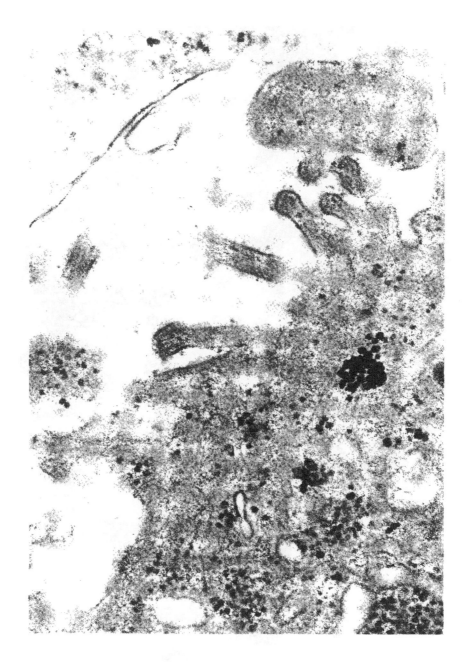

FIGURE 82. Cylindrical forms of respiratory syncytial virus budding from an infected monkey kidney cell. × 83,700.

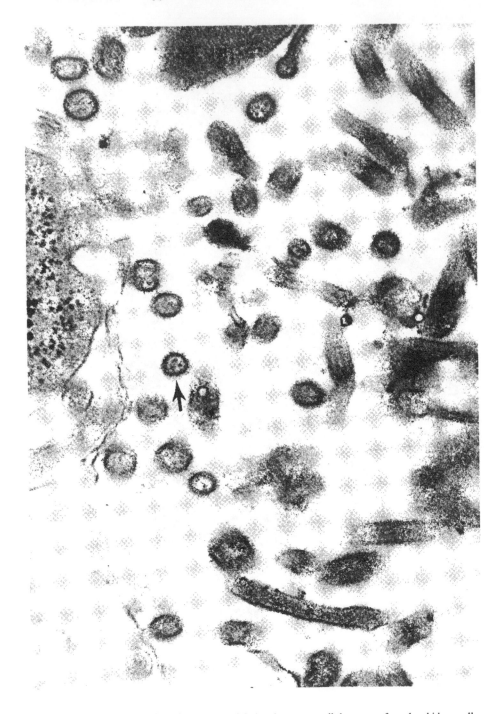

FIGURE 83.   Thin section of respiratory syncytial virus in an extracellular space of monkey kidney cells. Arrow points to a particle with surface projections and internal nucleocapsids seen in cross section. × 75,000.

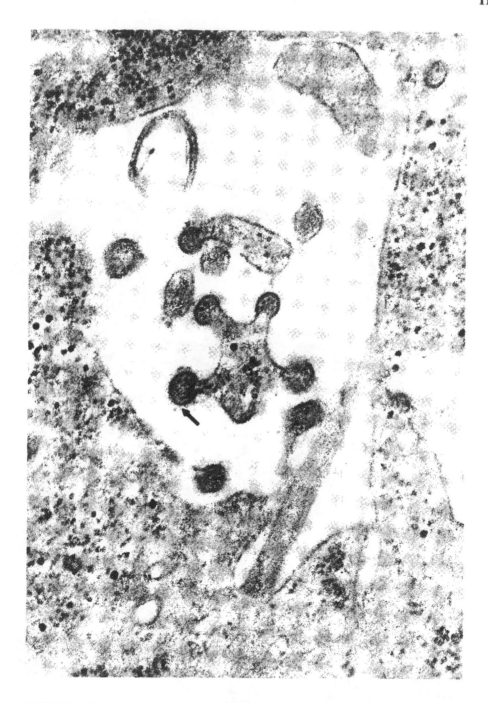

FIGURE 84.   Respiratory syncytial virus budding from cell membrane in an intracellular space of a monkey kidney cell. Projections (arrow) and internal nucleocapsids are evident. × 83,700.

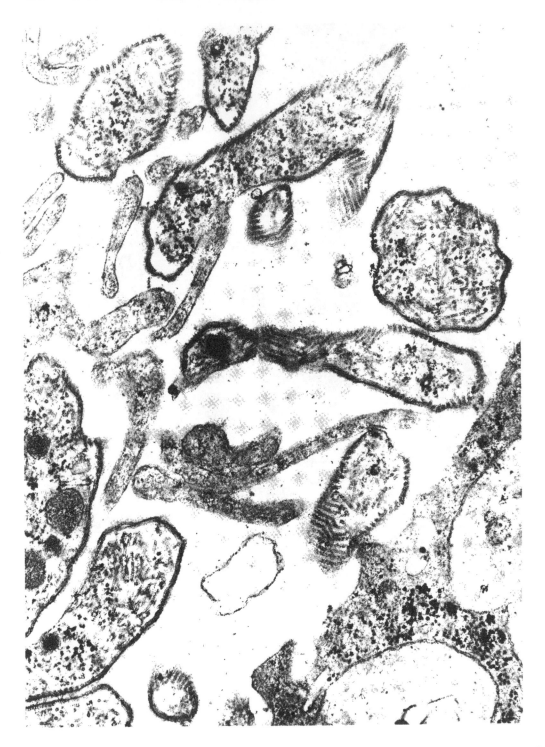

FIGURE 85.   Mumps virus in an extracellular space of monkey kidney cells. Nucleocapsids are seen in cross and longitudinal sections.   × 58,750. (Courtesy of Frederick Murphy, CDC.)

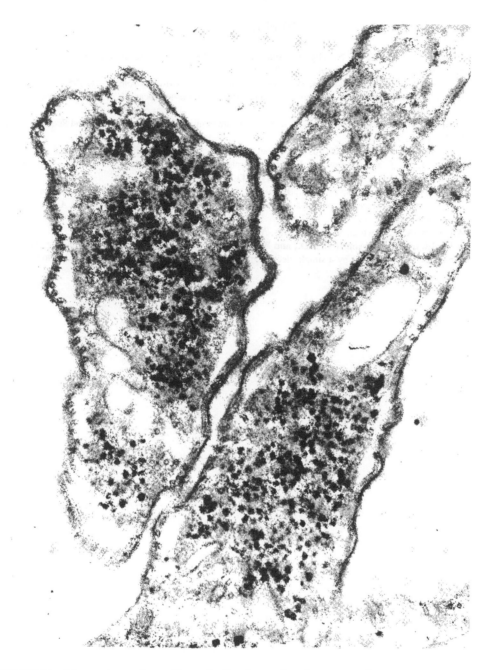

FIGURE 86. Mumps virus nucleocapsid strands aligned under the cell membrane of thin sectioned monkey kidney cells. Many are in cross section and fine material resembling surface projections is present on the outer membrane surface. × 102,750. (Courtesy of Frederick Murphy, CDC.)

REFERENCES

1. **Bloth, B., Espmark, A., Norrby, E., and Gard, S.,** The ultrastructure of respiratory syncytial (RS) virus, *Arch. Gesamte Virusforsch.*, 13, 582, 1963.
2. **Choppin, P. W. and Stoeckenius, W.,** The morphology of SV5 virus, *Virology*, 23, 195, 1964.
3. **Finch, J. T. and Gibbs, A. J.,** Observations on the structure of the nucleocapsids of some paramyxoviruses, *J. Gen. Virol.*, 6, 141, 1970.
4. **Horne, R. W. and Waterson, A. P.,** A helical structure in mumps, Newcastle disease and Sendai viruses, *J. Mol. Biol.*, 2, 75, 1960.
5. **Kingsbury, D. W.,** Paramyxovirus replication, *Curr. Top. Microbiol. Immunol.*, 59, 1, 1972.
6. **Norrby, E. C. J. and Magnusson, P.,** Some morphological characteristics of the internal component of measles virus, *Arch. Gesamte Virusforsch.*, 17, 443, 1965.
7. **Vorkunova, G. K., Pashova, V. A., Klimenko, S. M., Gushchin, B. V., and Bukrinskaya, A. G.,** The properties of intracellular paramyxovirus ribonucleoprotein in a nonpermissive system, *Arch. Gesamte Virusforsch.*, 46, 44, 1974.
8. **Waterson, A. P.,** Two kinds of myxovirus, *Nature (London)*, 193, 1163, 1962.
9. **Zakstelskaya, L. Y., Almeida, J. D., and Bradstreet, C. M. P.,** The morphological characterization of respiratory syncytial virus by a simple electron microscopy technique, *Acta Virol.*, 11, 420, 1967.

Chapter 14

CORONAVIRIDAE

Coronaviruses (family Coronaviridae) of importance to humans are recovered primarily from adults with upper respiratory tract illness. However, some reports indicate that these viruses can be involved in exacerbation of chronic lower respiratory tract disease as well. Coronaviruses have also been recovered or visualized from a host of lower animals, including murine, porcine, avian, bovine, canine, feline, and equine species. They cause a wide spectrum of illnesses in these animals, ranging from hepatitis in mice to diarrhea in calves. Strains that infect lower animals are not thought to cause disease in humans.

The human coronaviruses include at least three distinct but antigenically related prototype strains: B814, 229E, and OC43. Several uncharacterized subtypes have also been isolated from the human respiratory tract. Coronaviruses are fastidious in their growth requirements in the laboratory and may be divided into two groups, namely, those which can be isolated only in organ culture and those which can be isolated in cell culture. B814 and OC43 coronaviruses were originally isolated in human embryonic tracheal organ cultures, and 229E was originally isolated in human embryonic kidney cell culture. Subsequently, OC43 has been adapted to growth in suckling mouse brain and 229E in WI-38 fibroblasts and other cell cultures.

Coronavirus OC43 possesses a hemagglutinin. Although a hemagglutinin has not yet been demonstrated in other human strains thus far isolated, certain other animal strains possess a hemagglutinin. Coronaviridae infectivity is labile to pH, heat, and treatment with lipid solvents.

Coronaviruses are generally spherical, with some pleomorphism. They range from about 80 to 130 nm or more in diameter. Examination of coronavirus preparations by negative stain EM, however, usually reveals moderately pleomorphic particles covered with projections, known as peplomers (Figure 87). These projections are quite distinctive and give virions the appearance of a solar corona (hence the name coronavirus). The projections have a narrow base, and are 12 to 24 nm long. Their ends are club-shaped with an approximate diameter of about 10 nm. Symmetry to their arrangements has not yet been noted, although when one views them end-on, hexagonally arrayed projections are sometimes seen. The envelope can sometimes be seen when stain penetrates partially disrupted particles.

The genomic RNA of coronaviruses is of positive polarity and is infectious. Virions do not contain an RNA-dependent polymerase. Coronaviruses have a helical nucleocapsid but its exact structure has not been adequately defined. It has not been seen beneath the plasma membrane and cytoplasmic inclusions of nucleocapsid have not been detected. Budding does not occur from the plasma membrane. Virus multiplication occurs only in the cytoplasm. Particles mature by budding through the membrane of the cisternae of the endoplasmic reticulum and Golgi apparatus and may be found in cytoplasmic inclusions (Figure 88). Mature particles have an outer envelope about 8 nm thick surrounding an electron dense shell and a central zone enclosing a loosely wound helical ribonucleoprotein.

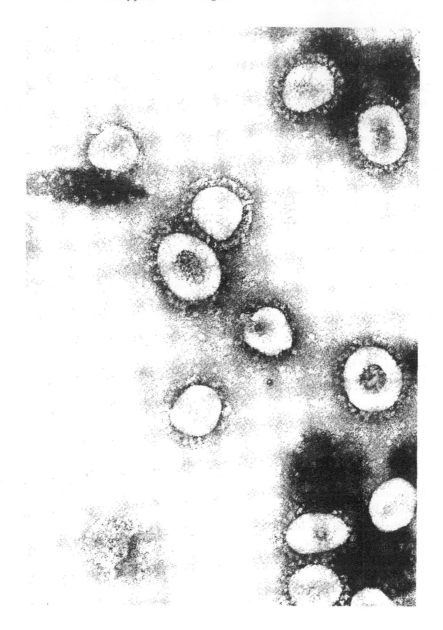

FIGURE 87.   Coronavirus OC43 negatively stained with UA. Virions are roughly circular and average about 100 nm in diameter. Particles have unique club-shaped peplomers which give the appearance of a solar corona. × 127,800.

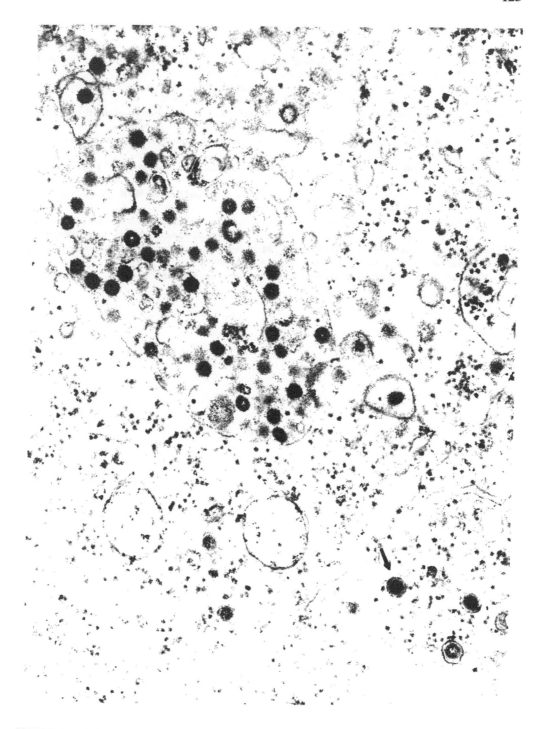

FIGURE 88. Calf diarrhea coronavirus in suckling mouse brain. Particles tend to accumulate in cytoplasmic inclusions. Complete particles with electron dense centers (arrow) are seen in the lower right of this picture. × 48,006. (Courtesy of Harold Kaye, CDC.)

REFERENCES

1. **Almeida, J. D., Berry, D. M., Cunningham, C. H., Hamre, D., Hosftad, M. S., Mallucci, L., McIntosh, K., and Tyrrell, D. A. J.,** Coronaviruses, *Nature (London),* 220, 650, 1968.
2. **Becker, W. B., McIntosh, K., Dees, J. H., and Chanock, R. M.,** Morphogenesis of avian infectious bronchitis and related human virus (strain 229E), *J. Virol.,* 1, 1019, 1967.
3. **Berry, D. M., Cruishank, J. G., Chu, H. P., and Wells, R. J. H.,** The structure of infectious bronchitis virus, *Virology,* 23, 403, 1964.
4. **Bingham, R. W. and Almeida, J. D.,** Studies on the structure of a coronavirus-avian infectious bronchitis virus, *J. Gen. Virol.,* 36, 495, 1977.
5. **Hamre, D., Kindig, D. A., and Mann, J.,** Growth and intracellular development of a new respiratory virus, *J. Gen. Virol.,* 1, 180, 1967.
6. **Hierholzer, J. C., Palmer, E. L., Whitfield, S. G., Kaye, H. S., and Dowdle, W. R.,** Protein composition of coronavirus OC43, *Virology,* 48, 526, 1972.
7. **Kapikian, A. Z.,** Coronaviruses, in *Diagnostic Procedures for Viral and Rickettsial Infections,* 4th Ed., Lennette, E. H. and Schmidt, N. J., Eds., American Public Health Association, New York, 1969, 931.
8. **Kennedy, D. A. and Johnson-Lussenburg, C. M.,** Isolation and morphology of the internal component of human coronavirus, strain 229E, *Intervirology,* 6, 197, 1975.
9. **McIntosh, K.,** Coronaviruses: a comparative review, *Curr. Top. Microbiol. Immunol.,* 63, 85, 1974.

Chapter 15

RHABDOVIRIDAE

The rhabdoviruses comprise all known bullet-shaped viruses. This unique form was first described for vesicular stomatitis virus (VSV) of horses and cattle. The group now consists of VSV, rabies, and other bullet-shaped particles isolated from vertebrates, plants, and insects. Rabies infects all warm-blooded mammals and is generally a fatal disease.

The main structure features of VSV and rabies are the bullet shape, an envelope, and a helical ribonucleoprotein consisting of single stranded RNA and protein. VSV measures about 170 nm long and 70 nm wide (Figure 89). The nucleoprotein is in the form of a single helix of about 30 turns and capped by 4 or 5 turns of diminishing size at the bullet-shaped end. The nucleoprotein is tightly packed into an envelope which has evenly spaced surface projections. Rabies virus is more bacilliform than VSV, is generally larger (200 to 300 nm), and varies more in diameter (50 to 80 nm). These two viruses belong to different genera of Rhabdoviridae and are antigenically distinct.

The replication of rabies and VSV follow the pattern of other negative stranded RNA viruses. During replication matrices of ribonucleoprotein form masses in the cytoplasm. These masses appear as inclusions (Negri bodies) that can be identified by immunofluorescence and are diagnostic for rabies. The inclusions formed by rhabdoviruses consist of unorganized RNP strands. These strands organize into helical RNP during the budding process. Nucleoproteins are not structured at sites remote from cell membranes. In contrast, the inclusion formed by filoviruses consists of highly structured nucleocapsids and nucleocapsid components. Rhabdovirus nucleocapsid inclusions are not membrane bound and may merge to eventually fill the cytoplasm.

Virions then assemble on preexisting cytoplasmic or plasma membranes. When sectioned longitudinally rhabdoviruses appear bullet shaped, and round or elliptical when sectioned transversely or obliquely (Figure 90). All rhabdoviruses are assembled by budding from the host cell membrane. Figure 91 shows dog rabies virus budding from a Negri body.

FIGURE 89.   Negatively stained (PTA) preparation of the rhabdovirus, vesicular stomatitis virus. The bullet shape of the virus is obvious. Particles have a helically wound nucleocapsid within an envelope from which projections originate. × 225,225. (Courtesy of Frederick Murphy, CDC.)

FIGURE 90. Navaro virus, a rhabdovirus, in suckling mouse brain. Disk shaped particles are seen when particles are sectioned tangentially. × 73,000. (Courtesy of Frederick Murphy, CDC.)

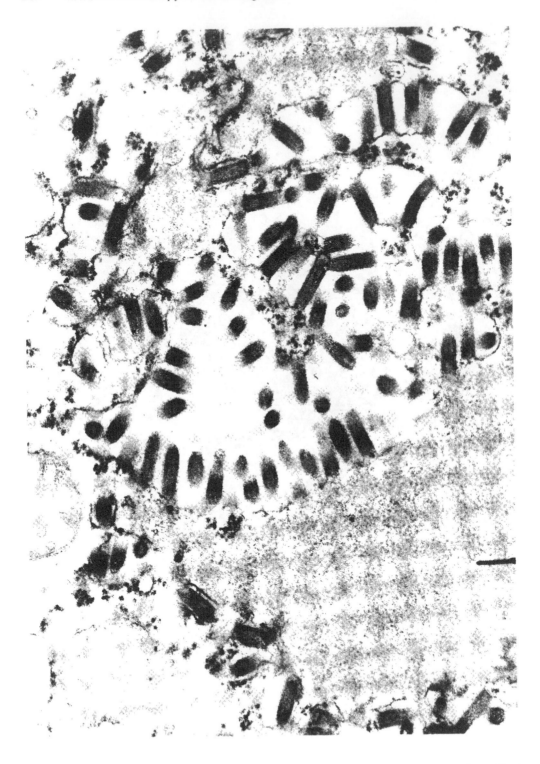

FIGURE 91.   Rabies virus in infected dog brain. Particles are budding into an intracellular space from a Negri body. × 82,350. (Courtesy of Alyne Harrison, CDC.)

REFERENCES

1. **Almeida, J. D., Howatson, A. F., Pinterci, L., and Fenje, P.,** Electron microscope observations on rabies virus by negative staining, *Virology,* 18, 147, 1962.
2. **Bergold, G. H. and Munz, K.,** Ultrastructure of Cocal, Indiana, and New Jersey serotypes of vesicular stomatitis virus, *J. Ultrastruct. Res.,* 17, 233, 1967.
3. **Bishop, D. H. L.,** Ed., *Rhabdoviruses,* CRC Press, Boca Raton, Fla., 1979.
4. **Bradish, C. J. and Kirkham, J. B.,** The morphology of vesicular stomatitis virus (Indiana C) derived from chick embryos or cultures of BHK 21/13 cells, *J. Gen. Microbiol.,* 44, 359, 1966.
5. **Howatson, A. F.,** Vesicular stomatitis and related viruses, *Adv. Virus Res.,* 16, 195, 1970.
6. **Howatson, A. F. and Whitmore, G. F.,** The development and structure of vesicular stomatitis virus, *Virology,* 16, 466, 1962.
7. **Murphy, F. A., Bauer, S., Harrison, A. K., and Winn, W. C.,** Comparative pathogenesis of rabies and rabies-like viruses. Viral infection and transit from inoculation site to the central nervous system, *Lab. Invest.,* 28, 361, 1973.

Chapter 16

BUNYAVIRIDAE

The virus family Bunyaviridae encompasses a large group of arthropod-borne viruses with morphologic and morphogenic similarities. The genus *Bunyavirus,* with Bunyamwera virus as the prototype, was designated to include all members of the Bunyamwera serological supergroup. There remained, however, other arboviruses which were morphologically similar to Bunyamwera virus but were antigenically unrelated to viruses of the *Bunyavirus* genus. Further serological, morphological, and biochemical analyses of these "Bunyavirus-like" viruses showed them to be distinctly different from viruses of the *Bunyavirus* genus. Subsequently, three other genera of Bunyaviridae were established — *Nairovirus* (prototype, Congo/Crimean hemorrhagic fever virus), *Phlebovirus* (prototype, Sandfly fever [Sicilian] virus, Sabin strain), and *Uukuvirus* (prototype — Uukuniemi virus). Viruses of this family are generally spherical, 80 to 120 nm in diameter, with surface projections 5 to 10 nm in length, which are anchored in a lipid bilayered envelope, and have a genome consisting of three segments of RNA (Figure 92). The surface structure of viruses within each Bunyaviridae genus is characteristic for that genus. The Bunyaviridae and Reoviridae are the only virus groups which have virus genera containing virusus which differ in morphology. Because of these differences, careful morphological studies of suspected Bunyaviridae may be used in considering preliminary generic assignment.

The ribonucleoprotein of Bunyaviridae is composed of three long helical circular strands about 2.5 nm in diameter. The ribonucleoprotein forms of LaCrosse have been designated as large (L), medium (M), and small (S). These forms were initially detected by double-staining gradient fractions with PTA and UA. Grids were prepared directly from high salt gradients by pseudoreplication onto UA then touched to a drop of PTA.

A fifth Bunyaviridae genus has been proposed to include a variety of antigenically distinct rodent virus isolates. These viruses are termed hemorrhagic fever with renal syndrome (HFRS) viruses. Hantaan virus, the etiologic agent of Korean hemorrhagic fever, has been designated the prototype virus. This is the most recent of the "elusive" viruses (Table 1, Introduction) to be detected by negative stain IEM. Initial studies of Hantaan virus revealed a bunyavirus-like morphology suggesting its identity as a member of the Bunyaviridae. In ultrastructural studies of glutaraldehyde-fixed viruses of the Bunyaviridae by negative stain EM, Hantaan virus and related viruses were found to have a surface structure composed of a grid-like pattern of morphological subunits not previously described for animal viruses (See Figure 17C, Chapter 2).

Bunyaviruses replicate in the cytoplasm of infected cells without involvement of a nucleocapsid precursor. The main site of virus maturation is in smooth endoplasmic membranes, mostly in association with the Golgi apparatus. Particles may bud into vesicles of the endoplasmic reticulum in the Golgi region and accumulate within cisternae of these organelles. Late in infection, cells become heavily vacuolated and some vacuoles are packed with virus. Virus is liberated from cells by fusion of intracellular vacuoles with the plasma membrane and virus egestion, or by cell disruption. There is no conclusive evidence that particles bud from the plasma membrane. Linear arrays of virions are often seen in intracellular spaces (Figure 93).

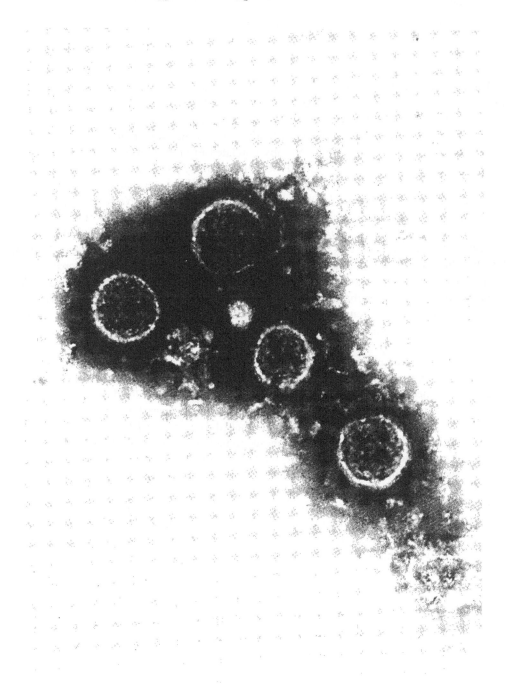

FIGURE 92.   Hantaan virus from the culture fluid of infected E-6 VERO cells negatively stained with PTA. Particles are penetrated by stain so that an envelope and surface projections are evident. × 195,325. (Courtesy of Joseph McCormick, CDC.)

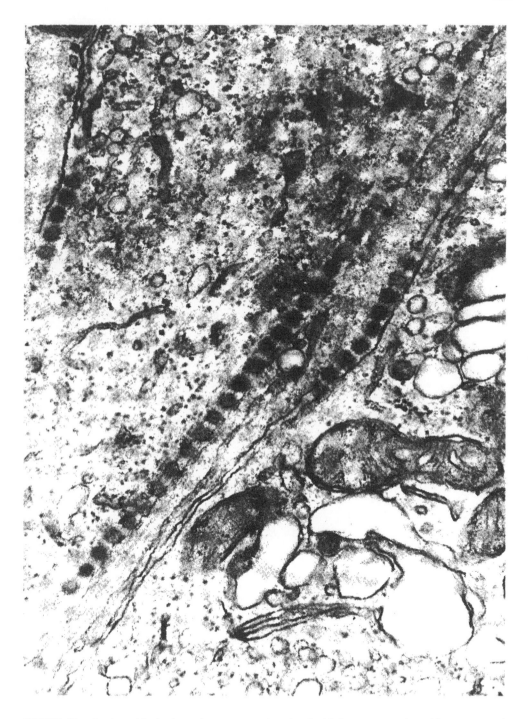

FIGURE 93.   Section of brain tissue from a mouse inoculated with Bunyamwera virus. Linear arrays of particles are seen in intracellular spaces. × 37,500. (Courtesy of Alyne Harrison, CDC.)

REFERENCES

1. **Brown, D. T., Waite, M. R. F., and Pfefferkorn, E. R.,** Morphology and morphogenesis of Sindbis virus as seen with freeze-etching techniques, *J. Virol.,* 10, 524, 1972.
2. **Enzmann, P. J. and Weiland, F.,** Studies on the morphology of alphaviruses, *Virology,* 95, 501, 1979.
3. **Horzinek, M. C.,** The structure of togaviruses, *Prog. Med. Virol.,* 16, 109, 1973.
4. **Jenson, A. B., Melnick, J. L., Edwards, M. R., and Whitney, E.,** California group arboviruses: electron microscopic studies, *Exp. Mol. Pathol.,* 9, 44, 1968.
5. **Murphy, F. A., Harrison, A. K., Gary, G. W., Jr., Whitfield, S. G., and Forrester, F. T.,** St. Louis encephalitis virus infection of mice. Electron microscopic studies of central nervous system, *Lab. Invest.,* 19, 652, 1968.
6. **Murphy, F. A., Harrison, A. K., and Whitfield, S. G.,** Bunyaviridae: morphologic and morphogenetic similarities of Bunyamwera serologic supergroup viruses and several other arthropod-borne viruses, *Intervirology,* 1, 297, 1973.
7. **Murphy, F. A., Harrison, A. K., and Tzianabos, T.,** Electron microscopic observations of mouse brain infected with Bunyamwera group arboviruses, *J. Virol.,* 1, 1315, 1968.
8. **Norrby, E.,** Rubella virus, *Virol. Mongr.,* 7, 115, 1969.
9. **Obijeski, J. F., Bishop, D. H. L., Palmer, E. L., and Murphy, F. A.,** Segmented genome and nucleocapsid of LaCrosse virus, *J. Virol.,* 20, 664, 1976.
10. **Von Bonsdorf, C-H. and Harrison, S. C.,** Sindbis virus glycoproteins form a regular icosahedral surface lattice, *J. Virol.,* 16, 141, 1975.
11. **Von Bonsdorf, C-H. and Pettersson, R.,** Surface structure of Uukuniemi virus, *J. Virol.,* 16, 1296, 1975.

Chapter 17

ARENAVIRIDAE

The name arenavirus is derived from the appearance of host cell ribosomes, which are seen as fine electron dense granules (arena = sand) within virus particles by thin section EM (Figure 94). The ribosomes are not seen by negative stain EM.

There are currently at least ten known viruses of the family Arenaviridae which fall into three groups: lymphocytic choriomeningitis virus (LCM); the Tacaribe complex which includes Junin (Argentinian hemorrhagic fever), Machupo (Bolivian hemorrhagic fever), Amapari, Tacaribe, Pichinde, Parana, Latino, and Tamiami viruses; and the Lassa virus group which includes Lassa virus, Moepia, Mobalaand, and Ippy viruses. Arenaviruses share a common group antigen but do not cross neutralize. Lassa virus and other arenaviruses which cause hemorrhagic fever are extremely hazardous and are currently studied in only a few laboratories with special safety precautions. The only known arenaviruses found in the U.S. are LCM, which may cause encephalitis in man, and Tamiami virus which is not known to be pathogenic for man.

All arenaviruses have the same morphology. Particles are enveloped, spherical to pleomorphic, and measure 50 to 300 nm in diameter. Viruses have a dense lipid bilayer envelope with surface projections (Figure 95). The projections are 10 nm long and are club-shaped. The core of arenaviruses contains several pieces of single-stranded RNA, two of which are virus-specific. Together with protein, this RNA forms circular ribonucleoprotein of at least two size modes of 650 and 1300 nm in length. The core also contains host cell ribosomal RNA and protein. The Arenaviridae and some members of the Bunyaviridae are the only viruses known to have circular, segmented ribonucleoprotein as nucleocapsids.

Arenaviruses mature by budding from the plasma membrane and membrane changes are detectable at the site of virus formation. There is also an alteration of the distribution of ribosomes. Aggregates of ribosomes may be dispersed throughout the cytoplasm. The cytoplasm also contains large inclusion bodies composed of masses of ribosomes in a matrix of virus-specific protein (Figure 96). A schematic of Arenavirus morphogenesis is shown in Figure 97. Late in infection, large areas of membrane may become dense and assimilation of host-cell ribosomes into virions can be seen.

Johnston Atoll virus, an unclassified arbovirus, closely resembles arenaviruses in morphology. Budding particles appear to accumulate ribosomes and particles contain electron dense structures which look like ribosomes (Figure 98). However, this virus does not appear to produce characteristic arenavirus cytoplasmic inclusion bodies. It also does not react with antisera to any of the known arenaviruses by fluorescent antibody tests.

Mycoplasma (Figure 99) may sometimes be confused with arenaviruses. These organisms are pleomorphic and some are about the same size and shape as arenaviruses. Mycoplasma may also contain internal electron dense bodies resembling ribosomes.

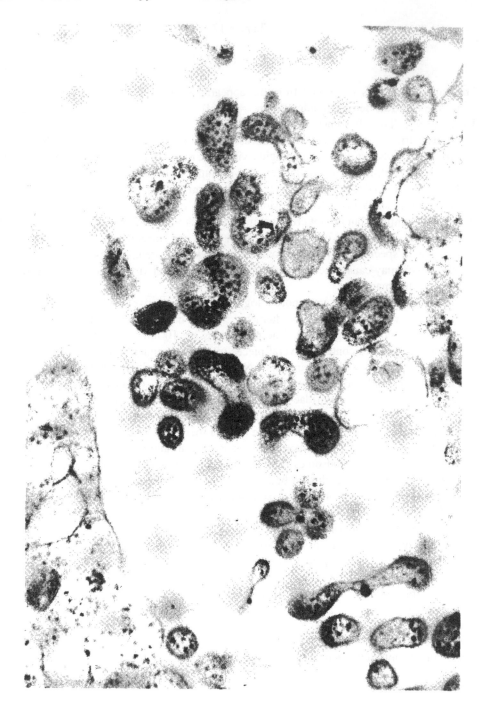

FIGURE 94.   Thin section of LCM virus in VERO cells. Particles are pleomorphic, variable in diameter, and have internal electron dense ribosomes. × 83,500. (Courtesy Sylvia Whitfield, CDC.)

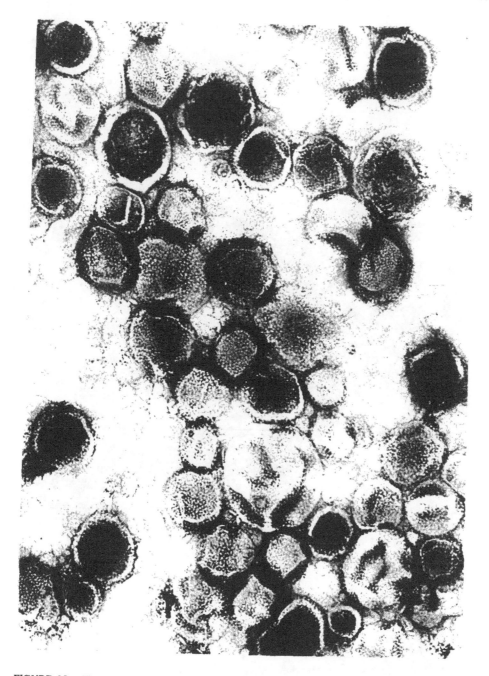

FIGURE 95. The arenavirus, Tacaribe, negatively stained with UA. Particles are pleomorphic and vary in size from 50 to 300nm in diameter. Surface projections are sometimes viewed head on and appear in hexagonal arrays. × 115,200.

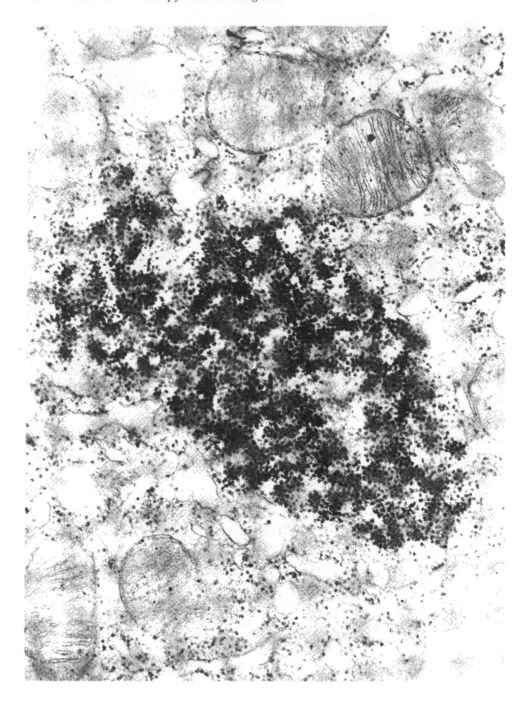

FIGURE 96.   Arenavirus inclusion in the cytoplasm of a monkey kidney cell. The inclusion consists of electron dense material and aggregations of cellular ribosomes. × 50,160. (Courtesy of Sylvia Whitfield, CDC.)

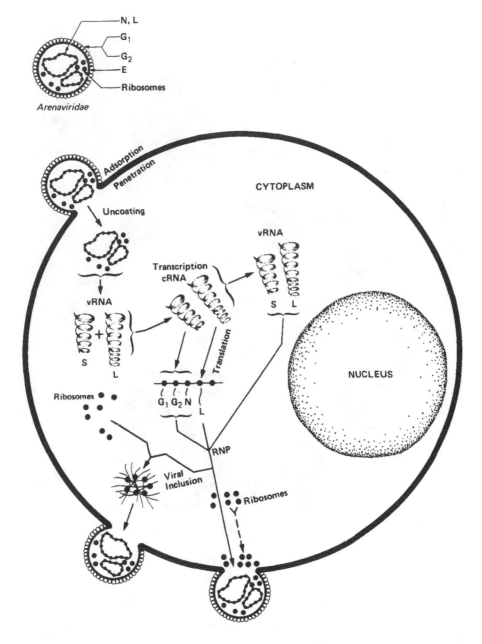

FIGURE 97. Schematic drawing showing an overview of events occurring during the replication of viruses of the Arenaviridae.

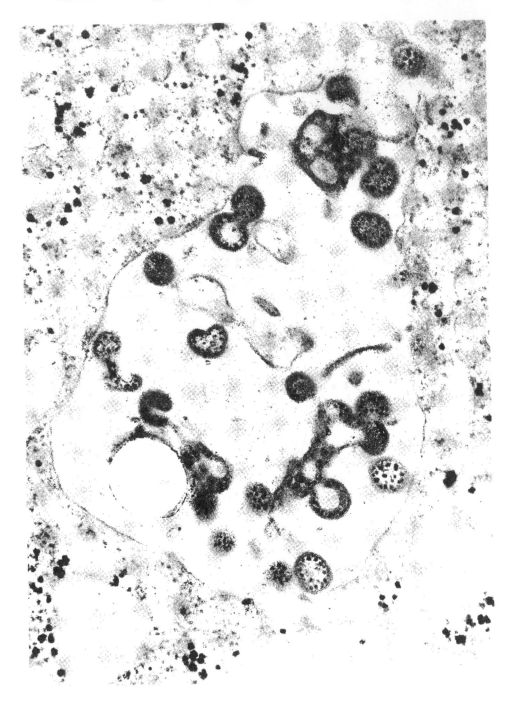

FIGURE 98.   Johnston Atoll virus, an unclassified arbovirus, in an intracellular space of an infected Hep-2 cell. Particles are roughly circular and have internal electron dense bodies resembling ribosomes. × 83,500.

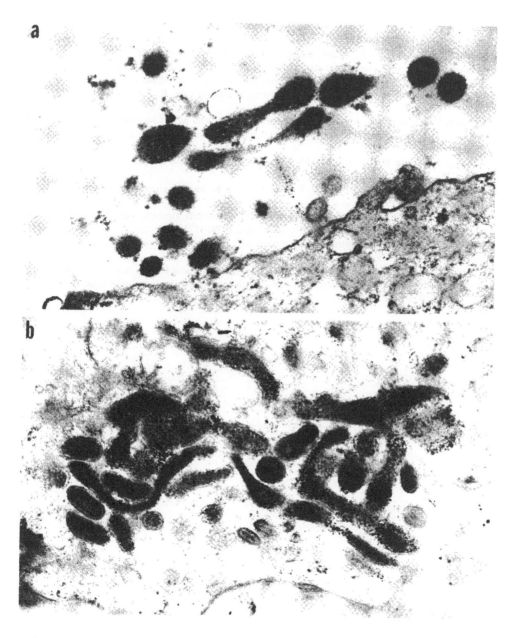

FIGURE 99. Mycoplasma in tissue culture cells. (A) Extracellular particles and (B) intracellular particles. Mycoplasma are generally pleomorphic and are often granular. × 74,450. (Courtesy of Alyne Harrison, CDC.)

REFERENCES

1. **Dalton, A. J., Rowe, W. P., Smith, G. H., Wilsnack, R. E., and Pugh, W. E.,** Morphological and cytochemical studies on lymphocytic choriomeningitis virus, *J. Virol.,* 2, 1465, 1968.
2. **Mannweiler, K. and Lehmann-Grube, F.,** Electron microscopy of LCM virus-infected L cells, in *Lymphocytic Choriomeningitis Virus and Other Arenavirus,* Lehmann-Grube, F., Ed., Springer Verlag, Berlin, 1973, 37.
3. **Murphy, F. A., Webb, P. A., Johnson, K. M., and Whitfield, S. G.,** Morphological comparision of Machupo with lymphocytic choriomeningitis virus: basis for a new taxonomic group, *J. Virol.,* 4, 535, 1969.
4. **Murphy, F. A. and Whitfield, S. G.,** Morphology and morphogenesis of arenaviruses, *Bull. WHO,* 52, 409, 1975.
5. **Palmer, E. L., Obijeski, J. F., Webb, P., and Johnson, K. M.,** The circular, segmented nucleocapsid of an arenavirus — Tacaribe virus, *J. Gen. Virol.,* 36, 541, 1977.
6. **Rawls, W. E. and Leung, W. C.,** Arenaviruses, in *Comprehensive Virology,* Vol. 14, Fraenkel-Conrat, H. and Wagner, R. R., Eds., Plenum Press, New York, 1979, 157.
7. **World Health Organization,** International symposium on arenavirus infections of public health importance, *Bull. WHO,* 52, 4, 1975.
8. **Spier, R. W., Wood, O., Liebheber, H., and Buckley, S. M.,** Lassa fever, a new virus disease of man from West Africa. IV. Electron microscopy of Vero cell cultures infected with Lassa virus, *Am. J. Trop. Med. Hyg.,* 19, 692, 1970.

Chapter 18

FILOVIRIDAE

The family Filoviridae (Filo = filament or thread) is comprised of Marburg virus and Ebola virus. The Marburg agent, which first caused serious hemorrhagic fever illness in a group of workers handling tissue from Ugandan African green monkeys, in Marburg, Germany, is a large rod-like virus frequently seen with one end curled in the forms of a 6 or a 9 (Figure 100). Negative staining reveals a helical ribonucleoprotein surrounded by an envelope covered with surface projections. The coiling of the nucleoprotein appears as a relatively wide axial channel (See Figure 21, Chapter 2). Bizarre branching forms are commonly seen by negative stain EM.

Ebola virus was the cause of an outbreak of hemorrhagic fever in the Sudan and Zaire in 1976. It was found to be serologically distinct from the Marburg agent, but had the same morphology. A second Ebola virus isolation was made from a human during an outbreak of hemorrhagic fever in Sudan in 1979. These viruses are very long rods and may be several microns in length (Figure 101). The length is highly variable. Branching is common, but particles have a fairly constant diameter of about 70 to 80 nm. Marburg and Ebola are negative stranded RNA viruses. Nucleocapsids form and accumulate in the cytoplasm. These appear as large inclusions. The viral internal constituents can be seen as cylinders in an amorphous matrix (Figures 102 and 103). Envelopes are added at the cell membranes, and surface projections are inserted as particles bud from the infected cells. During the budding process nucleocapsids may orient in any plane from perpendicular to parallel to cell membranes.

There have been no documented imported cases of Marburg or Ebola into the U.S. The diseases have thus far been confined to Africa except for the initial importation of Marburg into Germany. However, the extensive use of imported primary monkey cells makes it important that virologists be aware of these viruses. Both Marburg and Ebola viruses can possibly be spread by person to person contact.

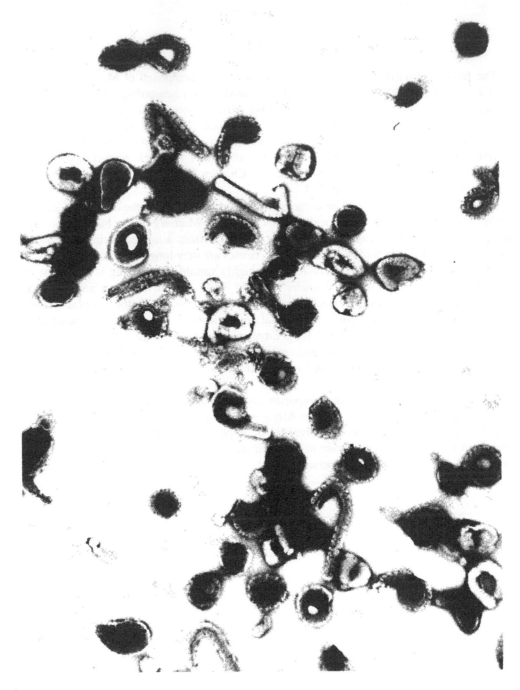

FIGURE 100.   Glutaraldehyde fixed Marburg virus purified from monkey kidney cells. This preparation has many "6" and "9" forms of the virus. UA stain, × 40,320. (Courtesy of Russell Regnery, CDC.)

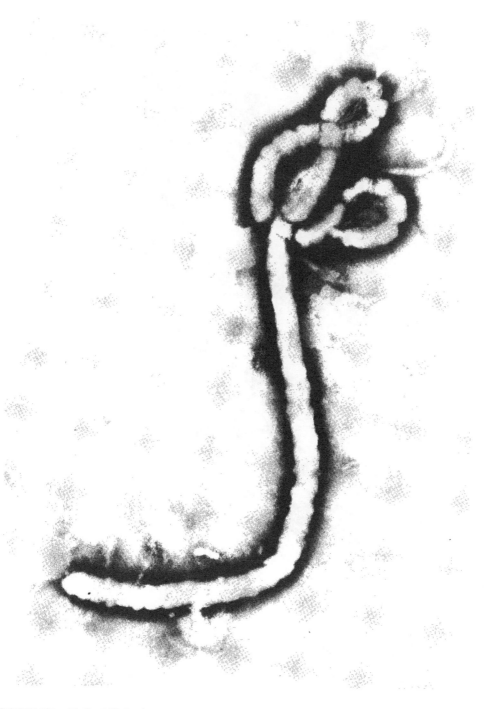

FIGURE 101.   Unfixed Ebola virus stained with PTA.   × 84,400. (Courtesy of Frederick Murphy, CDC.)

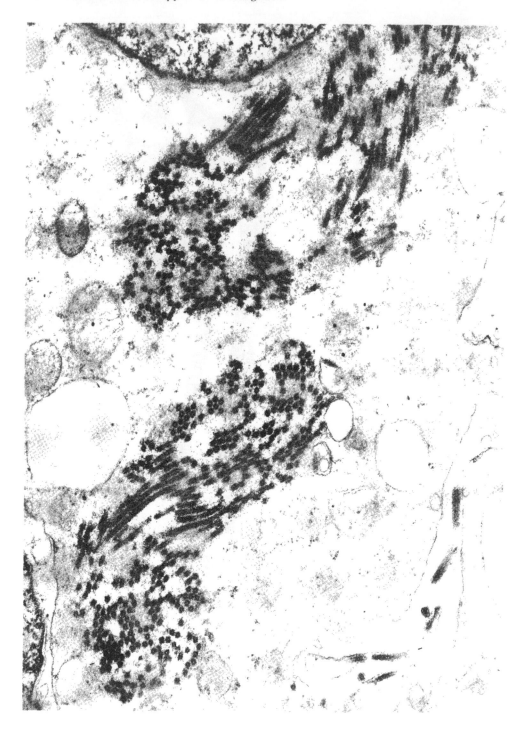

FIGURE 102.   Low magnification of Ebola virus inclusion in the cytoplasm of infected VERO cells. Particles are sectioned both tangentially and horizontally. × 31,617. (Courtesy of Sylvia Whitfield, CDC.)

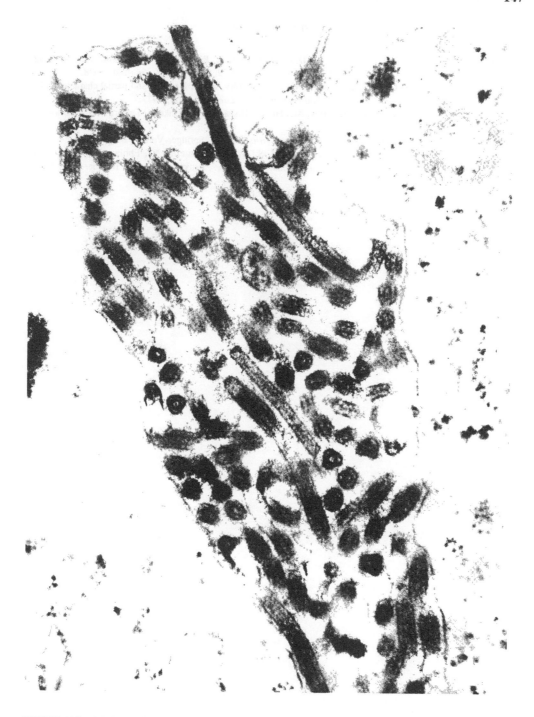

FIGURE 103.   Marburg virus in an extracellular space of monkey kidney cells. Virus particles are very long but have a fairly constant diameter. Virus was isolated directly from human liver. × 75,880. (Courtesy of Frederick Murphy, CDC.)

REFERENCES

1. **Ellis, D. S., Simpson, D. I. H., Francis, D. P., Knobloch, J., Bowen, E. T. W., Lolik, P., and Deng, I. M.,** Ultrastructure of Ebola virus particles in human liver, *J. Clin. Pathol.,* 31, 201, 1978.
2. **Johnson, K. M., Webb, P. A., Lange, J. V., and Murphy, F. A.,** Isolation and partial characterization of a new virus causing acute haemorrhagic fever in Zaire, *Lancet,* 1, 569, 1977.
3. **Malherbe, H. and Strickland-Cholmley, M.,** Human Disease from monkey (Marburg virus), *Lancet,* 1, 1434, 1968.
4. **Martini, G. A. and Siegert, R., Eds.,** *Marburg Virus Disease,* Springer-Verlag, Berlin, 1971.
5. **Murphy, F. A., Van der Groen, G., Whitfield, S. G., and Lange, J. V.,** Ebola and Marburg virus morphology and taxonomy, in *Ebola Virus Haemorrhagic Fever,* Pattyn, S. R., Ed., Elsevier, New York, 1978, 61.
6. **Siegert, R.,** Marburg virus, *Virol. Monogr.,* 11, 97, 1972.

Chapter 19

PARVOVIRIDAE

The parvoviruses are the smallest known viruses which infect vertebrates. The average particle size is about 22 nm in diameter. The symmetry of parvoviruses is icosahedral but the small size of the particle has hindered definitive detailed structural analysis of the capsid by currently available techniques. Some data indicate that these viruses are composed of at least 32 capsomeres. Viruses have single stranded DNA genomes which are either positive or negative DNA strands. The parvoviruses are divided into two groups: (1) the adenovirus-associated viruses (AAV), which are nonproductive and require adenoviruses for replication, and (2) a group of productive viruses such as latent rat virus H-1 and feline panleukopenia viruses, which can replicate autonomously. Whether AAV causes disease has been obscured by the requirement for the presence of adenoviruses for replication.

Parvovirus B 19 has been detected in the sera of patients with aplastic anemia during sickle cell crisis. Viruses can be easily detected in the serum by direct negative stain EM (Figure 104). This was the first human disease conclusively associated with parvoviruses. A similar, or probably the same, parvovirus is thought to cause erythema infectiosum (5th disease) in children. Particles described as "parvovirus-like" have also been detected in stools from persons with nonbacterial gastroenteritis.

Electron microscopy of ultra-thin sections has shown that parvoviruses replicate within the nucleus where virus may be seen in crystalline arrays. Late in infection particles are seen within membrane bound vesicles in the cytoplasm, or associated with endoplasmic reticulum. Figure 105 shows parvovirus H-1 in Chang liver cells. Empty particles are associated with ring-shaped structures, and empty and full (arrow) particles are dispersed throughout the nucleus.

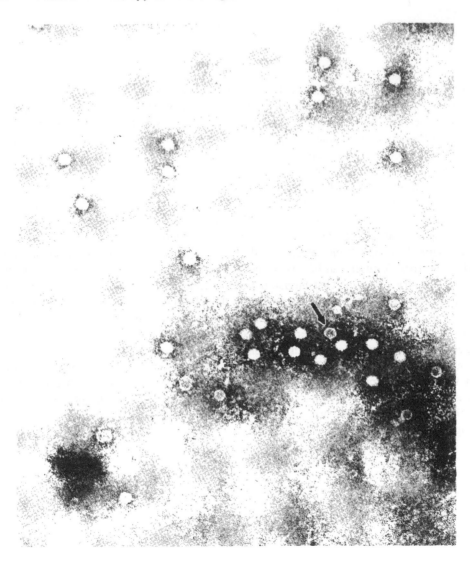

FIGURE 104.   Parvovirus in serum of a patient with acute red cell aplasia and hereditary hemolytic anemia. Particles are very small, icosahedral in shape, and relatively uniform in size. Empty forms are sometimes seen (arrow). UA stain × 156,260.

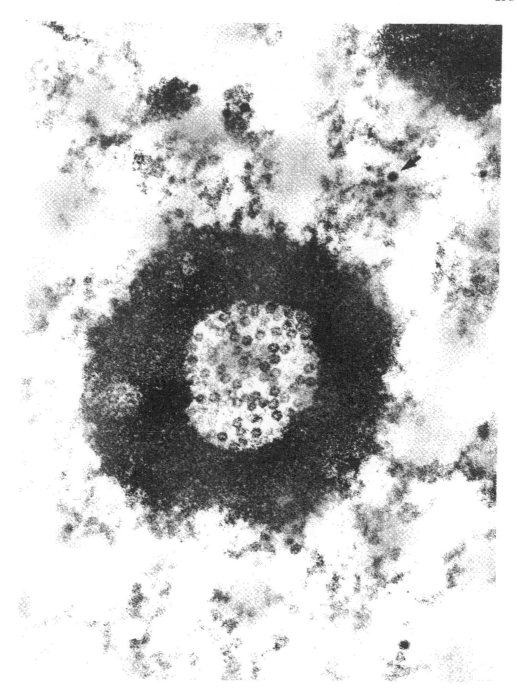

FIGURE 105. Parvovirus H-1 in Chang liver cells. Empty particles are within a ring shaped inclusion. Both empty and full (arrow) particles are dispersed throughout the cytoplasm. × 109,550.

REFERENCES

1. **Chandra, S. and Toolan, H. W.,** Electron microscopy of the H-1 virus. I. Morphology of the virus and a possible virus-host relationship, *J. Natl. Cancer Inst.,* 27, 1405, 1961.
2. **Chorba, T., Coccia, P., Holman, R. C., Tattersall, P., Anderson, L. J., Sudman, J., Young, N. S., Kurczynski, E., Saarinen, A. M., Moir, R., Lawrence, D., Jason, J. M., and Evatt, B.,** Parvovirus B-19 infection: the cause of epidemic aplasia crisis and erythema infectiosum (Fifth Disease), *J. Inf. Dis.,* 154, 383, 1986.
3. **Hoggan, M. D.,** Adenovirus associated viruses, *Prog. Med. Virol.,* 12, 211, 1970.
4. **Hoggan, M. D.,** Small DNA viruses, in *Comparative Virology,* Maramorosch, K. and Kurstak, E., Eds., Academic Press, New York, 1971, 43.
5. **Karasaki, S.,** Size and ultrastructure of the H-viruses as determined with the use of specific antibodies, *J. Ultrastruct. Res.,* 16, 109, 1966.
6. **Mayor, H. D., Jamison, R. M., Jordan, L. E., and Melnick, J. L.,** Structure and composition of a small particle prepared from a simian adenovirus, *J. Bacteriol.,* 90, 235, 1965.
7. **Saarinen, U. M., Chorba, T. L., Tattersall, P., Young, N. S., Anderson, L. J., Palmer, E., and Coccia, P. F.,** Human parvovirus B-19-induced epidemic acute red cell aplasia in patients with hereditary hemolytic anemia, *Blood,* 67, 1411, 1986.
8. **Serjeant, G. R., Topley, J. M., and Mason, K.,** Outbreak of aplastic crisis in sickle cell anemia associated with parvovirus-like agent, *Lancet,* 2, 595, 1981.
9. **Toolan, H. W., Saunders, E. L., Green, E. L., and Fabrizio, D. P. A.,** Further studies on the electron microscopy of the H-1 virus, *Virology,* 22, 286, 1964.

Chapter 20

HERPESVIRIDAE

Herpes simplex virus was the first animal virus to be studied by negative stain EM. Two basic types of herpesvirus particles have been described. One is a naked nucleocapsid and the other a nucleocapsid surrounded by an envelope. The capsid is composed of capsomeres arrayed as an icosahedron such that each capsid is composed of 162 hollow prismatic capsomeres. The diameter of the nucleocapsid is 100 to 105 nm. The nucleocapsid forms in the nucleus and the envelope is derived from the nuclear membrane. Enveloped particles vary in diameter from about 120 to over 200 nm. A core, consisting of a fibrillar spool on which DNA is wrapped, is anchored to the underside of the capsid shell. Enveloped particles usually contain only one nucleocapsid, but it is not uncommon to see two or three inside one envelope. Enveloped particles vary in size when seen in negative stain preparations, but are usually 150 to 180 nm in diameter (Figure 106). When enveloped particles are not penetrated by stain they are difficult to distinguish from cell membrane material. Enveloped particles also have regularly spaced surface projections which are about 10 nm in length. The significance of these projections is not well understood.

In micrographs of thin sections, a herpesvirus capsid measuring about 100 nm can be clearly resolved (Figure 107). It encloses a double-stranded DNA genome. There is also some evidence that herpesviruses have a middle capsid layer just beneath the capsid and an inner capsid about 45 nm in diameter which encloses the core. However, substantiation of such discrete structures is not yet unequivocal. For resolution of these components as discrete structures, a gentle method of controlled degradation of herpes virions must be found. When particles mature at the nuclear membrane, they are enclosed by a membrane (Figure 108).

The herpesviruses of importance in medicine are herpes simplex virus types 1 and 2, varicella zoster virus (chickenpox and shingles), Epstein-Barr virus, cytomegalovirus (CMV), human B cell lymphotropic virus, and herpes simiae (herpes B). EM has made an important contribution to the rapid diagnosis of herpesviruses from vesicular fluids, urine, and brain biopsy specimens (thin section). Attempts to type these viruses by IEM have been less successful but current efforts at IEM indicate that the problems involved in typing studies may soon be overcome. CMV can be detected in urine by direct EM using the pseudoreplica technique.

As shown in the schematic (Figure 109), herpesvirus enters the cell by a process similar to pinocytosis or by fusion with the cell. The particle is uncoated and DNA is liberated. The DNA is transcribed to form an RNA which migrates to ribosomes where it directs the synthesis of viral structural and nonstructural proteins. Some of these migrate to the nucleus, and others bind to cellular membranes. The structural proteins in the nucleus aggregate with DNA and form the capsid. The capsid then becomes enveloped at the nuclear membrane. The exact process by which herpesvirus is released from the cell is not clear. However, particles appear to accumulate between the inner and outer lamella of the nuclear membrane or in the endoplasmic reticulum. The endoplasmic reticulum may connect the perinuclear spaces with extracellular fluid and provide a pathway for virus egress. Also, portions of endoplasmic reticulum may break off and form vesicles which are released to extracellular fluids.

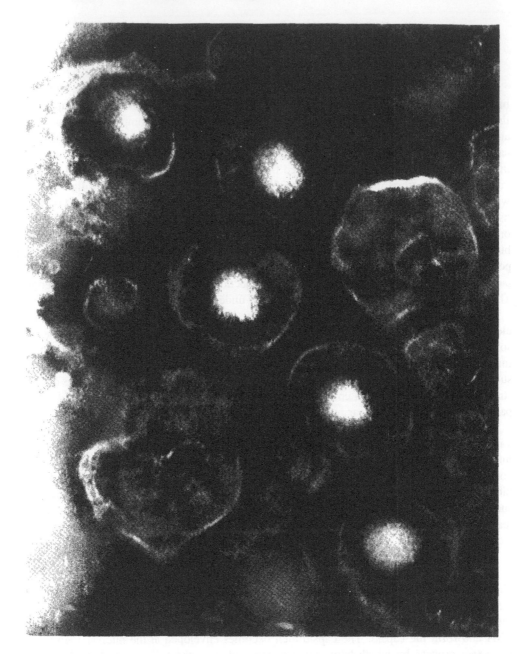

FIGURE 106.   Herpes simplex virus negatively stained with UA. An icosahedral nucleocapsid is surrounded by an envelope with indistinct surface projections. Virus purified from human lung fibroblast cells. × 153,360.

FIGURE 107. Herpes simplex virus nucleocapsids in the nucleus of a MA104 cell. Particles are hexagonal in shape and many have electron dense cores. × 70,000.

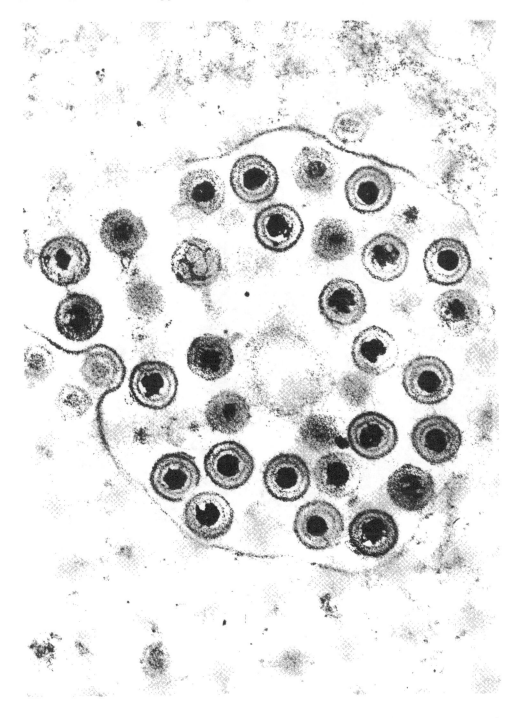

FIGURE 108. Complete herpes simplex virus particles in a cytoplasmic vacuole of an infected MA104 cell. Core, capsid, and envelope are evident. × 85,000.

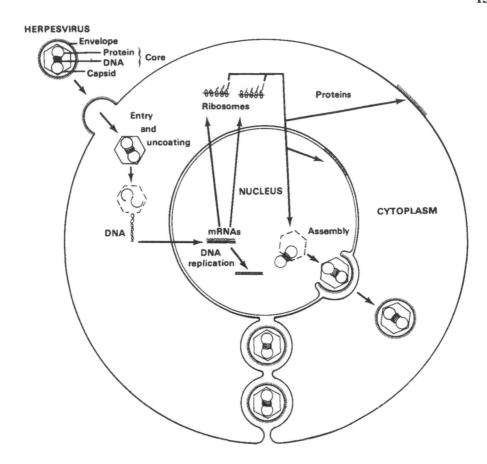

FIGURE 109. Schematic showing an overview of the replication of viruses of the Herpesviridae. Virus entry, uncoating, replication of intermediates, assembly into virions, and egress from the infected cell are depicted.

REFERENCES

1. **Almeida, J. D., Howatson, A. F., and Williams, M.,** Morphology of varicella (chickenpox) virus, *Virology,* 16, 353, 1962.
2. **Epstein, M. A. and Achong, B. G.,** Morphology of the virus and of virus-induced cytopathologic changes, in *The Epstein-Barr Virus,* Epstein, M. A. and Achong, B. G., Eds., Springer-Verlag, Berlin, 1979, 23.
3. **Horne, R. W. and Wildy, P.,** Symmetry in virus architecture, *Virology,* 15, 348, 1961.
4. **Morgan, C., Ellison, S. A., Rose, H. M., and Moore, D. H.,** Structure and development of viruses as observed in the electron microscope. I. Herpes simplex virus, *J. Exp. Med.,* 100, 195, 1954.
5. **Nii, S. and Yasuda, I.,** Detection of viral cores having toroid structures in eight herpesviruses, *Biken J.,* 18, 41, 1975.
6. **Palmer, E. L., Martin, M. L., and Gary, G. W., Jr.,** The ultrastructure of disrupted herpesvirus nucleocapsids, *Virology,* 65, 260, 1975.
7. **Watson, D. H.,** The structure of animal viruses in relation to their biological functions, *Symp. Soc. Gen. Microbiol.,* 18, 207, 1968.
8. **Watson, D. H.,** Morphology, in *The Herpesviruses,* Kaplan, A. S., Ed., Academic Press, New York, 1973, 27.
9. **Wildy, P., Russell, W. C., and Horne, R. W.,** The morphology of herpes virus, *Virology,* 12, 204, 1960.

Chapter 21

ADENOVIRIDAE

Adenoviruses have been recovered from patients with a variety of clinical syndromes such as febrile pharyngitis, epidemic keratoconjunctivitis, acute respiratory disease, and gastroenteritis. There are currently more than 42 known serotypes which infect mammals. The viruses induce latent infections in lymphoid tissue and are readily activated. The initial isolation of adenovirus was made from human adenoid tissue grown as a cell culture in which cells degenerated after prolonged incubation. More recently, new serotypes of adenovirus have been recovered from stools of persons with nonbacterial gastroenteritis, and a causal relationship has been established. The mammalian adenoviruses are the genus *Mastadenovirus*.

Negatively stained adenovirus particles are 70 to 90 nm in diameter (Figure 110). Particles consist of a naked nucleocapsid composed of 252 capsomeres arranged in icosahedral symmetry. There are 240 hexons (group antigen) and 12 pentons (type specific antigen). Each penton has a fiber attached which is difficult to see by EM either because these structures are seldom extended or become disattached easily. The capsid surrounds a genome of dsDNA.

Adenoviruses are present in large numbers in cell culture when CPE is pronounced. When present in stools, adenovirus is also usually in large numbers so direct EM is a useful method to group adenovirus isolates. IEM has also been used to type several common adenoviruses and the serum in agar method has been used to detect canine adenovirus.

The site of adenovirus maturation is the nucleus. Nuclei develop characteristic basophilic inclusion bodies composed of accumulated unassembled viral components (Figure 111). Particles are adsorbed to and penetrate susceptible cells. The capsid disintegrates and the core containing DNA migrates to the nucleus, or DNA enters the nucleus through a membrane pore. DNA replication and mRNA transcription occur in the nucleus and viral proteins are synthesized on polyribosomes in the cytoplasm. These are transported to the nucleus, associate with DNA, and are assembled into virions (Figure 112).

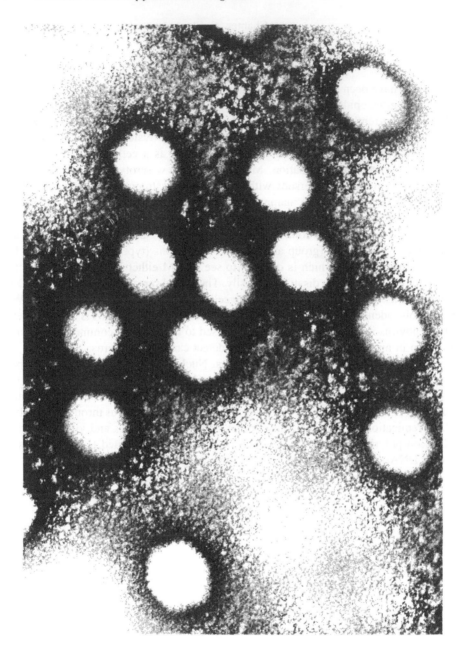

FIGURE 110.   Adenovirus negatively stained with PTA. Particles have a hexagonal outline and capsomeres have a distinctive array. Virus propagated in monkey kidney cells. × 297,000.

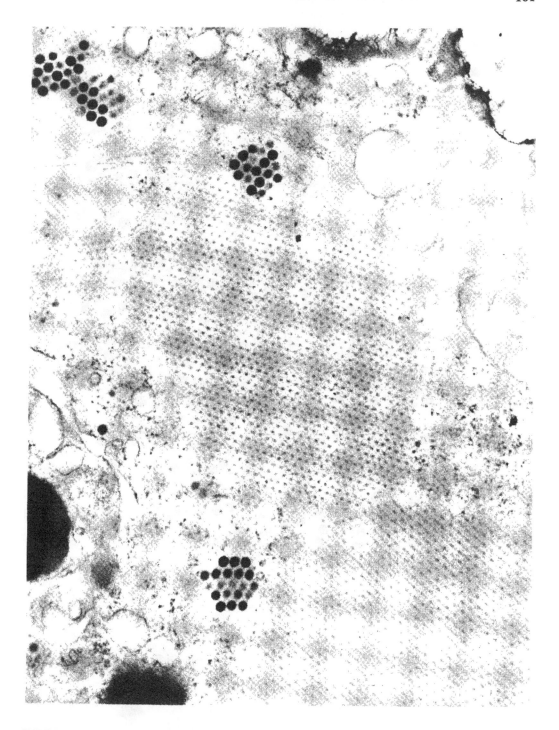

FIGURE 111. Inclusions in the nucleus of a lymphocyte infected with adenovirus. Virus was isolated from the lymphocytes of a patient with AIDS. Arrays of adenovirus particles can be seen at the periphery of inclusions. × 41,073. (Courtesy of Alyne Harrison, CDC.)

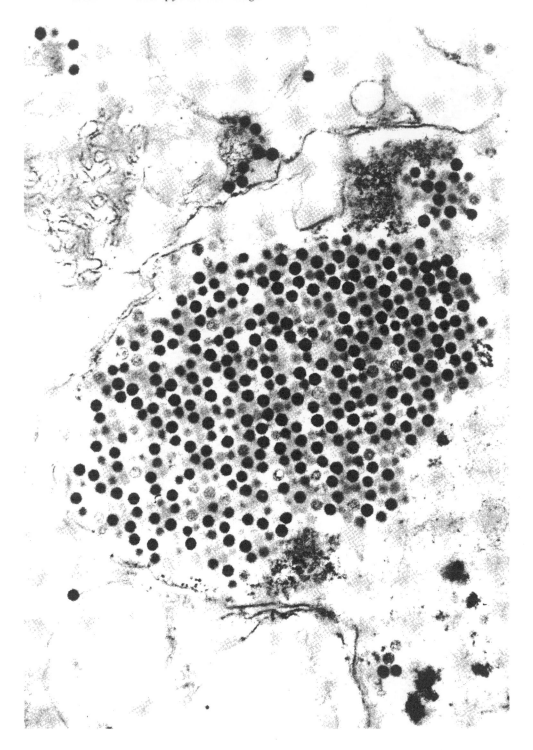

FIGURE 112.   Hexagonal arrays of adenovirus in the nucleus of a lymphocyte. Virus was isolated from the lymphocytes of a patient with AIDS. × 62,230. (Courtesy of Alyne Harrison, CDC.)

REFERENCES

1. **Hilleman, M. R. and Werner, J. R.,** Recovery of a new agent from patients with acute respiratory illness, *Proc. Soc. Exp. Biol. Med.,* 85, 183, 1954.
2. **Horne, R. W.,** The comparative structure of adenoviruses, *Ann. N. Y. Acad. Sci.,* 101, 475, 1962.
3. **Horne, R. W., Brenner, S., Waterson, A. P., and Wildy, P.,** The icosahedral form of an adenovirus, *J. Mol. Biol.,* 1, 84, 1959.
4. **Nermut, M. V.,** Fine structure of adenovirus type 5, *Virology,* 65, 480, 1975.
5. **Norrby, E.,** The structural and functional diversity of adenovirus capsid components, *J. Gen. Virol.,* 5, 221, 1969.
6. **Pereira, H. G. and Wrigley, N. G.,** In vitro reconstitution, hexon bonding and handedness of incomplete adenovirus capsid, *J. Mol. Biol.,* 85, 617, 1974.
7. **Smith, K. O., Gehle, W. D., and Trousdale, M. D.,** Architecture of the adenovirus capsid, *J. Bacteriol.,* 90, 254, 1965.
8. **Valentine, R. C. and Pereira, H. G.,** Antigens and structures of the adenovirus, *J. Mol. Biol.,* 13, 13, 1965.

Chapter 22

PAPOVAVIRIDAE

Viruses of the Papovaviridae include papilloma and polyoma viruses, and simian vacu-olating virus (SV40). The prefix "papova" is derived from papilloma, polyoma, and vac-uolating agent (SV-40). The name "papova" is composed of the first two letters of each of these virus groups. Polyoma infects mice and produces neoplasias of different types. A structurally similar virus, SV40, multiplies silently in primary monkey kidney cells. SV40-like viruses have been isolated from the brains of patients with progressive multifocal leukoencephalopathy (papovavirus JC) and then from urine of immunosupressed individuals (papovavirus BK). Over 70% of individuals tested have antibody to the SV40-like viruses.

Polyoma virions are naked icosahedrons about 45 nm in diameter composed of 72 cap-someres in a skew arrangement with T = 7 symmetry. The genome is a cyclic double stranded DNA. Papilloma virions are structurally similar to those of polyoma but are slightly larger (about 55 nm). These viruses produce papillomas or benign warts in mammals, including man. Papilloma viruses are easily recovered from extracts of warts and can be seen by direct EM of wart tissue. However, these viruses do not replicate in tissue culture. Human papilloma virus has a right-handed, or dextro, arrangement of capsomeres, whereas rabbit papilloma virus has a left-handed, or levo, capsomere arrangement. Capsomeres of the virus are easily seen by negative staining EM (Figure 113).

Viral DNA replication and assembly of polyoma virus capsids occurs in the nucleus (Figure 114 and 115) and virus release depends on cell disruption. Viral DNA replication occurs in supercoiled intermediates with several distinct proteins expressed from the same DNA segment. Tubular forms of the virus are also commonly seen in cells expressing virus.

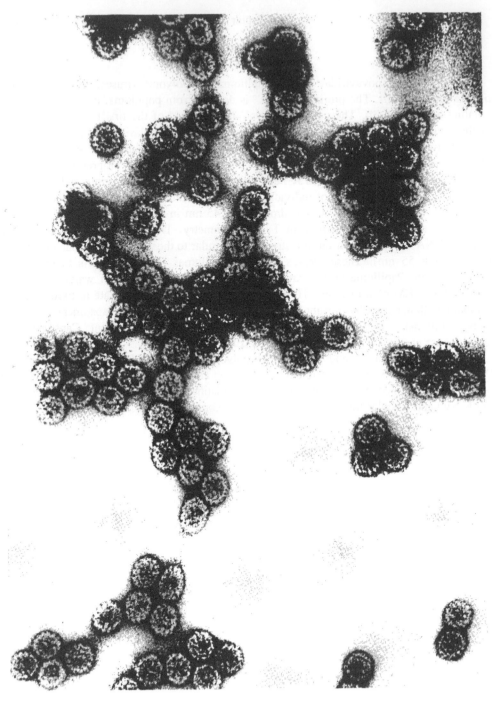

FIGURE 113.   Negative stain of purified polyoma virus. Capsomeres are arranged in a T = 7 skew pattern. UA stain × 127,800.

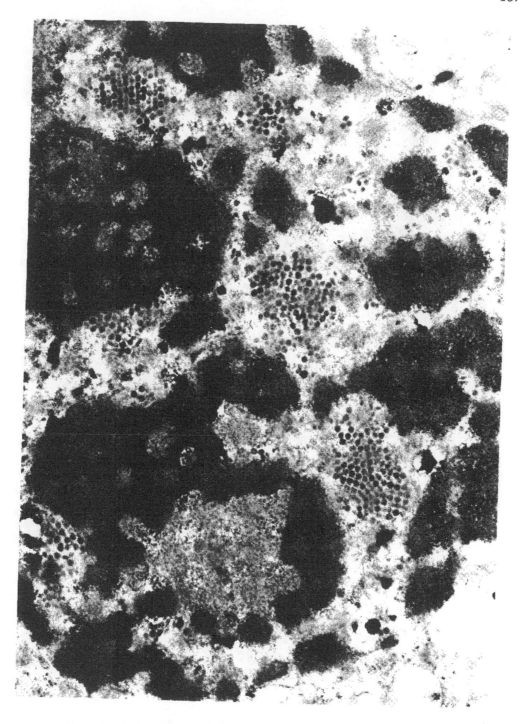

FIGURE 114. Arrays of papovavirus BK in a brain cell of a patient with PML resulting from AIDS. Particles are in hexagonal arrays. Papovaviruses can be distinguished by virus location (nucleus), size, and overall morphology. × 30,000.

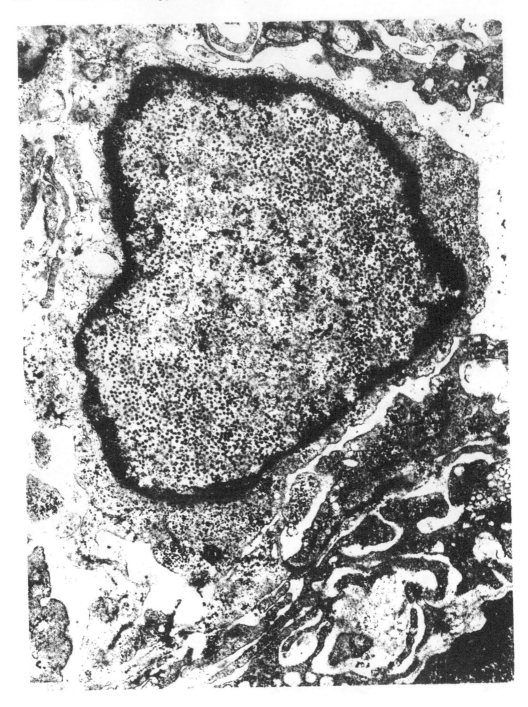

FIGURE 115.   Low magnification of a brain cell from a patient with PML resulting from AIDS. Numerous papovavirus particles can be seen in the cell nucleus.   × 13,500.

REFERENCES

1. **Almeida, J. A., Howatson, A. F., and Williams, M. G.,** Electron microscope study of human warts, sites of virus production and nature of the inclusion bodies, *J. Invest. Dermatol.,* 38, 337, 1962.
2. **Finch, J. T.,** The surface structure of polyoma virus, *J. Gen. Virol.,* 24, 359, 1974.
3. **Finch, J. T. and Klug, A.,** The structure of viruses of the papilloma-polyoma type. III. Structure of rabbit papilloma virus, *J. Mol. Biol.,* 13, 1, 1965.
4. **Howatson, A. F. and Crawford, L. V.,** Direct counting of the capsomers in polyoma and papilloma viruses, *Virology,* 21, 1, 1963.
5. **Kiselev, N. A. and Klug, A.,** The structure of viruses of the papilloma-polyoma type. V. Tubular variants built of pentamers, *J. Mol. Biol.,* 40, 155, 1969.
6. **Klug, A.,** The structure of viruses of the papilloma-polyoma type. II. Comments on other work, *J. Mol. Biol.,* 11, 424, 1965.
7. **Klug, A. and Finch, J. T.,** The structure of viruses of the papilloma-polyoma type. I. Human wart virus, *J. Mol. Biol.,* 11, 403, 1965.
8. **Klug, A. and Finch, J. T.,** The structure of viruses of the papilloma-polyoma type. IV. Tilting experiments and two side images, *J. Mol. Biol.,* 31, 1, 1968.

REFERENCES

Chapter 23

POXVIRIDAE

The poxviruses are the largest and most complex of vertebrate viruses. There are a number of strains which are pathogenic for different animal species. These are divided into six genera. Two of these genera *(Orthopoxvirus* and *Parapoxvirus)* are important as pathogens of humans. Member viruses in each genus cross-react extensively in most conventional serological tests, but only one antigen is common to all genera. This antigen is considered to be a nucleoprotein complex.

Variola virus, the causal agent of smallpox, is in the *Orthopoxvirus* genus. Since the beginning of recorded history smallpox has plagued entire populations. However, because there is no animal reservoir for variola virus and because of the effectiveness of vaccination, smallpox is now believed to have been globally eradicated. Only variola and molluscum contagiosum viruses are specific human pathogens. Other diseases such as Orf and cowpox are actually diseases of lower animals that may be transmitted to man. Another example is the simian poxvirus, Tanapox, which can naturally infect humans. It does not belong to the variola-vaccinia group, but is serologically related to Yaba virus of African monkeys located in a wide area of central Africa. Other poxviruses, such as camelpox and ectromelia, cause eruptive skin lesion diseases in lower animals. Except for viruses of the Parapoxvirus genus, all poxviruses are brick-shaped with a surface pattern resembling whorled filaments, and are about 225 × 300 nm in size (Figure 116). The parapoxviruses (Orf and others) appear more ovoid with a regular spiraled filamentous surface pattern, and average only about 150 × 200 nm in size (Figure 117). The internal structure of all poxviruses studied has been found to be essentially identical.

Two types of poxvirus particles, originally called elementary bodies, are seen in tissue culture or vesicular fluids: brick-shaped particles, termed C particles, with a wide, clearly defined envelope and an electron-dense center; and mulberry (M) particles which have a whorled surface filament pattern. The M particles usually have a central area which represents the structure of the core or nucleoid surrounding a double-stranded DNA genome and the lateral bodies, which are attached at each side of the nucleoid. The envelope pushes the bodies against the core so that the latter is dumbbell shaped. The lateral bodies and envelope can be removed by treating virions with detergent and 2-mercaptoethanol. The core then becomes expanded and brick-shaped.

Although the genome of poxviruses is double stranded DNA, the synthesis of viral components and the assembly of virions takes place within the cytoplasm of infected cells. Virus enters the cell by a process of engulfment. It is partially uncoated by cellular enzymes, then DNA is released by uncoating proteins derived from transcripts within the core. DNA synthesis proceeds, proteins are formed from late transcripts, and virus is formed in cytoplasmic "factories" or viral inclusions (Figure 118). The stages of poxvirus morphogenesis are shown in the schematic in Figure 119.

Immature particles are usually spherical with the genome folded into an electron-dense nucleoid. Lateral bodies differentiate and an envelope is laid down over these structures. Virus is released by cell lysis or through cell villi.

Vaccinia virus is currently being used in a number of laboratories as an expression vehicle for genetically engineered viral vaccines. The genes producing the proteins responsible for production of neutralizing antibody of a variety of viruses, including rabies, Lassa fever, influenza, and HIV have been cloned. These proteins have been expressed in vitro after infection of cells with vaccinia virus containing genes of various viruses.

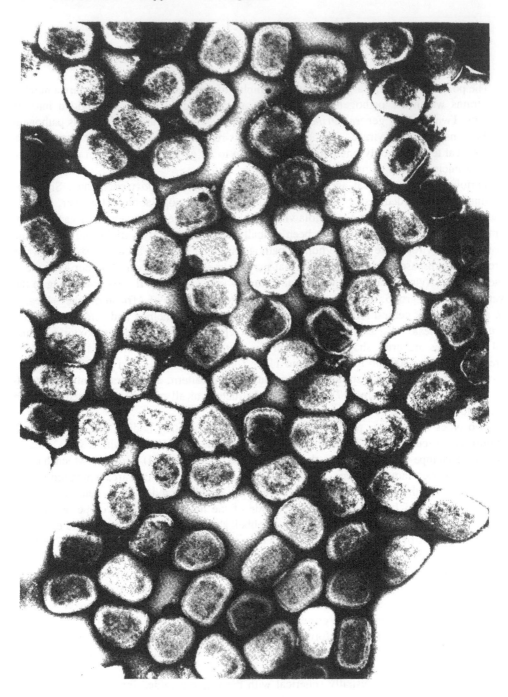

FIGURE 116.   Purified preparation of vaccinia virus negatively stained with UA. Most particles have a ''brick''
shape. × 50,000.

FIGURE 117. Orf virus negatively stained with PTA. Particle on left is unpenetrated by stain and has a whorled surface pattern. Particle on right is penetrated by stain such that the envelope is evident. × 124,875. (Courtesy of Patricia Bingham, CDC.)

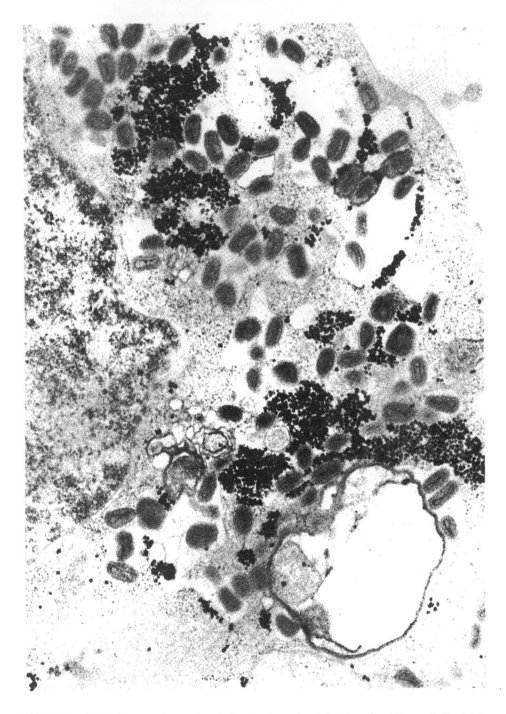

FIGURE 118.   Cotia virus, an orthopoxvirus, in the cytoplasm of an infected monkey kidney cell. Dumbbell pattern of nucleocapsid is evident. × 31,590. (Courtesy of Alyne Harrison, CDC.)

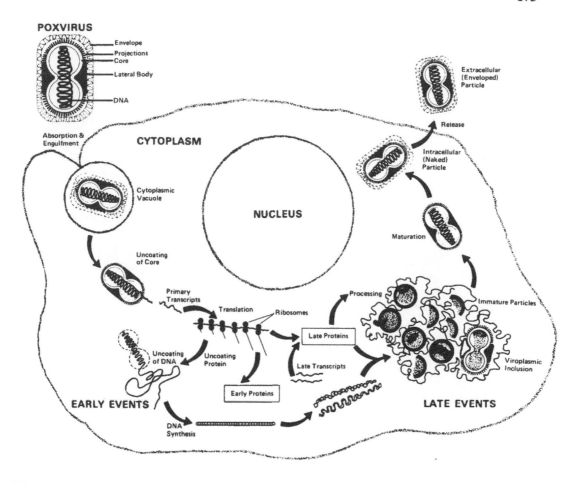

FIGURE 119. Schematic overview of the events occurring during the adsorption, uncoating, formation of intermediates, assembly of virions, and egress of poxvirus from a cell.

REFERENCES

1. **Dales, S. and Siminovitch, L.,** The development of vaccinia virus in Earle's L strain cells as examined by electron microscopy, *J. Biophys. Biochem. Cytol.*, 10, 475, 1961.
2. **Dales, S. J.,** The uptake and development of vaccinia virus in L strain cells followed with labeled viral deoxyribonucleic acid, *J. Cell Biol.*, 18, 51, 1963.
3. **Dales, S. J. and Mosbach, E. H.,** Vaccinia as a model for membrane biogenesis, *Virology*, 35, 564, 1968.
4. **Easterbrook, K. B.,** Controlled degradation of vaccinia virions in vitro: an electron microscopic study, *J. Ultrastruct. Res.*, 37, 132, 1966.
5. **Mackett, M. and Smith, G. L.,** Vaccinia virus expression vectors. *J. Gen. Virol.*, 67, 2067, 1986.
6. **Medzon, E. L. and Bauer, H.,** Structural features of vaccinia virus revealed by negative staining, *Virology*, 40, 860, 1970.
7. **Mitchiner, M. B.,** The envelope of vaccinia and orf viruses: an electron-cytochemical investigation, *J. Gen. Virol.*, 5, 211, 1969.
8. **Nagington, J., Newton, A. A., and Horne, R. W.,** The structure of orf virus, *Virology*, 23, 461, 1964.
9. **Noyes, W. F.,** The surface fine structure of vaccinia virus, *Virology*, 17, 282, 1962.
10. **Westwood, J. C. N., Harris, W. J., Zwartouw, H. T., Titmuss, D. H. J., and Appleyard, G.,** Studies on the structure of vaccinia virus, *J. Gen. Microbiol.*, 34, 67, 1964.

Chapter 24

HEPADNAVIRIDAE AND OTHER HEPATITIS VIRUSES

There are now thought to be at least three types of hepatitis virus. These are (1) Hepatitis A (HAV), or infectious hepatitis, usually transmitted through an intestinal-oral route; (2) Hepatitis B (HBV), or serum hepatitis, usually transmitted through infected blood or its products; and (3) Non A Non B (NANB) hepatitis which is the major etiologic agent of transfusion hepatitis in the U.S. The structure of HAV and HBV is known. NANB virus has not been isolated or identified.

Hepatitis A virus is a 27 nm naked icosahedron with a single stranded RNA genome (Figure 120). Morphologically it appears to be a picornavirus and has been designated as Enterovirus 72. However, the classification of HAV is currently being reevaluated on the basis of recent RNA sequencing analysis. Virus can be passaged in nonhuman primates and virus detected in stools by IEM. HAV also replicates in some cells. Multiplication occurs in the cytoplasm and is probably similar to that of picornaviruses. There are specific serological tests to detect HAV antibody in serum.

Hepatitis B is a DNA virus. The DNA is double stranded and cyclic. HBV is usually seen in sera in one of three forms. The infectious particle (Dane particle) is 42 nm in diameter with an electron dense core about 28 nm in diameter (Figure 121). This is the least common particle in sera of persons with HBV. The most common form seen in serum is the surface antigen (HBsAg) or Australia antigen (see Figure 23, Chapter 3). This form is about 22 nm in diameter and appears similar in morphology to the surface of Dane particles. These particles are noninfectious and contain no nucleic acid. Filamentous forms, also 22 nm in diameter and varying in length up to several hundred nm, are also commonly seen in serum. HBV is the prototype of the virus family, Hepadnaviridae.

The gene coding for the surface antigen has been cloned and the antigen expressed in bacteria, yeast, and tissue culture. The current vaccines for HBV are surface antigen purified from human plasma and a genetically engineered surface antigen expressed in yeast. Efforts are currently underway to test the effectiveness of the genetically engineered antigen. Figure 122 is a negative stain EM of HBsAg produced by hamster embryo cells into which the gene for surface antigen has been inserted. There are currently some culture systems for propagation of HBV, but the morphogenesis of the virus has not been adequately defined.

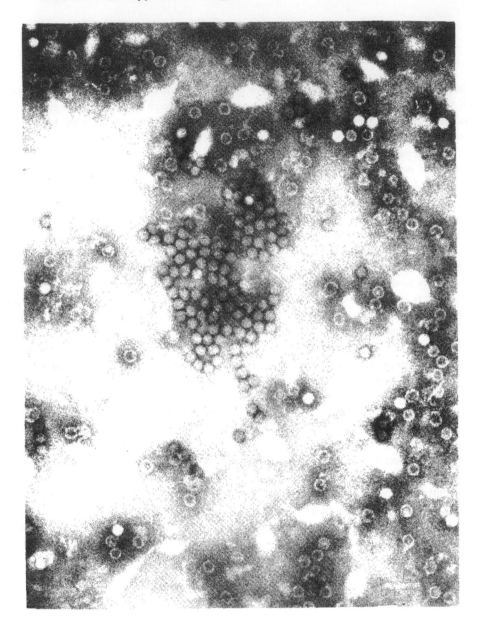

FIGURE 120.   Group of hepatitis A particles purified from infected cell culture fluid. Virus is hexagonal and about 27 nm in diameter. PTA stain × 108,420. (Courtesy of Jim Cook, CDC.)

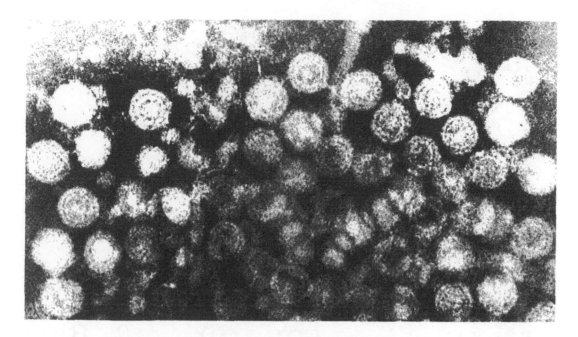

FIGURE 121.   Negatively stained hepatitis B virus. Particles purified from the stool of a patient with hepatitis B. Particles are generally uniformly round with a diameter of 42 nm. A core is also evident. PTA stain × 257,000. (Courtesy of Jim Cook, CDC.)

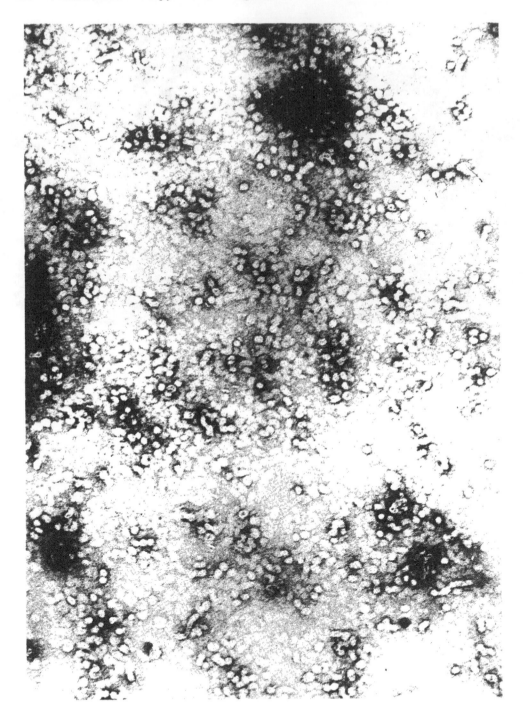

FIGURE 122.   Hepatitis B surface antigen purified from hamster embryo cell culture fluids and negatively stained with UA. The gene which codes for the surface antigen was artificially inserted into the cells. Purified particles are about 22 nm in diameter. Some are rod-shaped. × 83,700. (Courtesy of John Obijeski, Genentech, Inc.)

REFERENCES

1. **Almeida, J. D., Waterson, A. P., Trowell, J. M., and Neal, G.,** The finding of virus-like particles in two Australia-antigen-positive human livers, *Microbios,* 2, 6145, 1970.
2. **Bayer, M. E., Blumberg, B. S., and Werner, B.,** Particles associated with Australia antigen in the sera of patients with leukaemia, Down's syndrome and hepatitis, *Nature (London),* 218, 1057, 1968.
3. **Dane, D. S., Cameron, C. H., and Briggs, M.,** Virus-like particles in serum of patients with Australia-antigen-associated hepatitis, *Lancet,* 1, 695, 1970.
4. **Feinstone, S. M., Kapikian, A. Z., and Purcell, R. H.,** Hepatitis A: detection by immune electron microscopy of a virus-like antigen associated with acute illness, *Science,* 182, 1026, 1973.
5. **Feinstone, S. M., Moritsugu, Y., Shih, J. W. K., Gerin, J. L., and Purcell, R. H.,** Characterization of hepatitis A virus, in *Viral Hepatitis: A Contemporary Assessment of Etiology, Epidemiology, Pathogenesis and Prevention,* Vyas, G. N., Cohen, S. N., and Schmid, R., Eds., Franklin Institute Press, Philadelphia, 1978, 41.
6. **Gerin, G. L.,** Structure of hepatitis B antigen (HBAg), in *Mechanisms of Virus Disease,* Robinson, W. S. and Fox, C. R., Eds., Benjamin, Menlo Park, Calif., 1974, 215.
7. **Robinson, W. S. and Lutwick, L. I.,** The virus of hepatitis, type B., *N. Engl. J. Med.,* 295, 1168, 1976.
8. **Siegel, G. and Frosner, G. G.,** Characterization and classification of virus particles associated with hepatitis A. 1. Size, density and sedimentation, *J. Virol.,* 26, 40, 1978.

Chapter 25

SCANNING ELECTRON MICROSCOPY

An image seen by transmission EM is produced by electrons emitted from a heated filament. These electrons pass through a condenser lens system forming an electron beam which illuminates specimens. Some electrons are scattered or deflected by the specimen, but others pass through and produce an image on a fluorescent screen. The image can then be recorded on photographic emulsion sensitive to electrons. The resolution is very high with currently available transmission EMs (≥ 1 Å). An image seen by scanning EM (SEM) is produced by a small focus of electrons that sequentially scans the surface of the specimen. Physical effects caused by the incident beam as it scans the surface can be detected and displayed on a monitor. This image is a visual enlargement of the topography of the specimen being scanned.

In scanning EM the specimen is fixed, dehydrated in a graded series of alcohols, dried by critical point drying, and shadowed with metals. Virus structure, except for general shape, is not well defined by SEM. However, SEM does give a better overall view of the extent of infection with some viruses than does transmission electron microscopy (TEM). For instance, TEM of thin sections of T4 lymphocytes infected with HIV reveals virus at the plasma membrane or in intracellular spaces because only a section of a cell is viewed and virus is formed only at the plasma membrane. In contrast, HIV is readily seen on the surface of T4 lymphocytes by SEM because the face of a whole cell can be scanned. SEM photomicrographs of an HIV infected T4 lymphocyte are shown in Figure 123.

Numerous particles are seen attached to the cell. At high magnification particles can be seen budding from infected cells (Figures 124 and 125). SEM can be compared to TEM of HIV infected lymphocytes (see Figures 66 and 67, Chapter 11).

Figure 126 shows a SEM photomicrograph of monkey kidney cells infected with Ebola virus. By SEM the virus appears as a "snake-like" structure with no defined surface or internal structures as is seen with negative stain TEM.

It is clear that viruses cannot yet be identified by direct SEM. SEM is most useful for visualizing viruses which bud from the cell membrane and for checking cell cultures for mycoplasma.

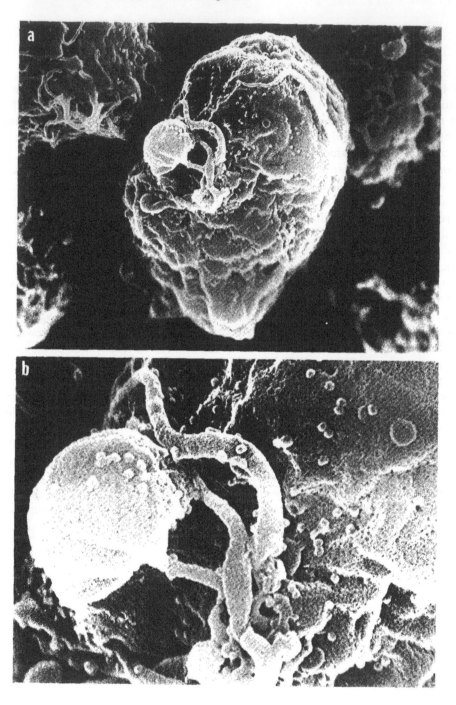

FIGURE 123.   Scanning EM of an H-9 T4 lymphocyte infected with HIV. At low magnification virus is seen as ''blebs'' on the cell surface. (A) Whole cell, × 15,100; (B) higher magnification of same cell, × 30,450. (Scanning EM courtesy of Cynthia Goldsmith, CDC.)

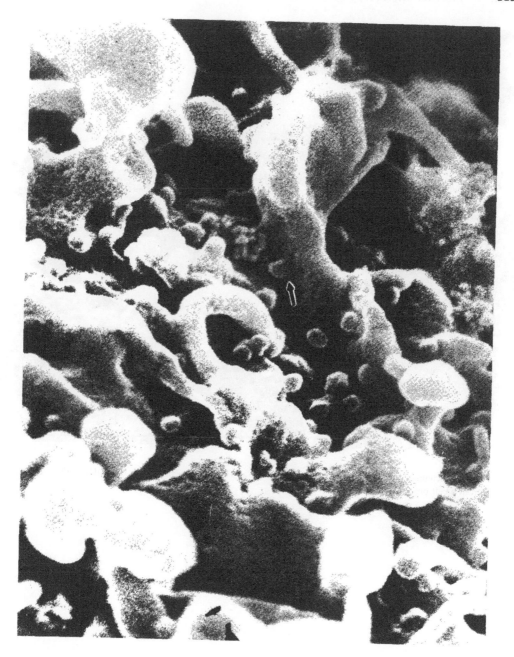

FIGURE 124.   Scanning EM of the surface of a T4 lymphocyte infected with HIV. Virus can be seen attached to or budding from (arrow) the cell membrane. × 75,000.

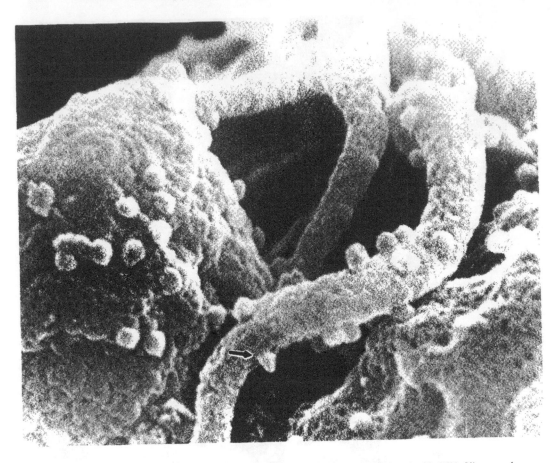

FIGURE 125.   Scanning EM of a portion of the surface of an H-9 lymphocyte infected with HIV. Virus can be seen budding from a cell process (arrow) × 75,000.

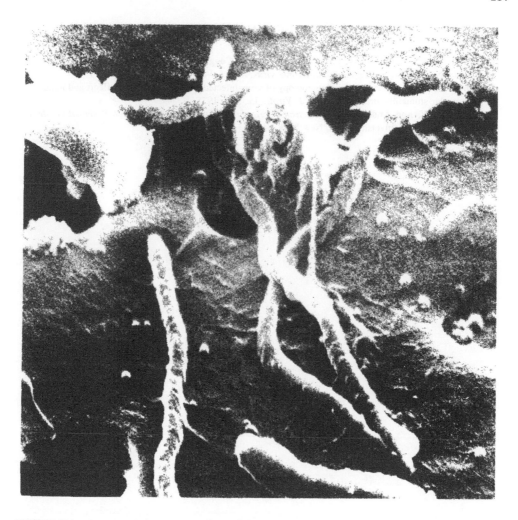

FIGURE 126. Scanning EM of Ebola virus isolated from infected E-6 VERO cells. The virus appears as a "snakelike" structure with no definitive surface structure. × 75,000.

REFERENCES

1. **Gonda, M. A., Charman, H. P., Walker, J. L., and Coggins, L.,** Scanning and transmission electron microscopic study of equine infectious anemia virus, *Am. J. Vet. Res.,* 39, 731, 1978.
2. **Holmes, K. V.,** Scanning electron microscopy of virus-infected cells. I. Cytopathic effects and maturation of vesicular stomatitis virus in L2 cells, *J. Virol.,* 15, 353, 1975.
3. **Doane, F. W. and Anderson, N.,** *Electron Microscopy in Diagnostic Virology: A Practical Guide and Atlas,* Cambridge University Press, New York, 1986.

INDEX

Printed and bound by CPI Group (UK) Ltd, Croydon, CR0 4YY

22/10/2024

01777632-0003